Preface

The study of inorganic reaction mechanisms, which dates from the 1860s, is by no means new. The majority of information has, however, been obtained since the 1950s. With the advent of instrumental methods of analysis and the development of methods for the observation of fast reactions, the field is one which has expanded rapidly.

As with most areas of chemistry which have received more than a passing share of the limelight in recent years, there has been a tendency to introduce the subject at school level. The purpose of this monograph is, hopefully, to provide some suitable information to facilitate this introduction. Obviously, in a text of this size it is not possible to provide even a remotely comprehensive survey of such a wide subject. Consequently, selection is inevitable and, furthermore, no one area can be treated in any great depth. The selection is, of course, personal but attention has been paid to the type of reactions encountered in current A-, OND and HND level courses. Some attempt has been made to illustrate the role of metal ions in a number of important organic and biochemical reactions. In addition, attention has been paid to clock reactions and the more recently discovered oscillatory reactions which are particularly suitable for generating interest and enthusiasm.

A number of experiments which have been found useful are included. In many cases an integrated exercise may be built around the experiment involving, for example, the synthesis and analysis of a compound followed subsequently by investigation of the kinetics and mechanism of one or more of its reactions. The majority of experiments quoted may readily be extended to form the basis of projects.

Safety
The experiments described in this monograph have been used successfully for a number of years, however, no guarantee of their safety can be given.

Nomenclature

ASE[1] nomenclature has been used in this monograph. The oxoacid and transition metal species are named in full, *eg* dioxochlorate(III) (chlorate(III)), tetraoxomanganate(VI) (manganate(VI)) *etc.* Where no ASE recommendations are available IUPAC[2] nomenclature has been used.

The most frequently mentioned species are dioxonitric(III) acid (nitrous acid), hydrogen trioxosulphate(IV) (hydrogen sulphite), tetraoxomanganate(VI) (permanganate), trioxosulphate(IV) (sulphite) and trioxothiosulphate(VI) (thiosulphate).

The following abbreviations are used:

aq	aquated, H_2O
k	rate constant
K	equilibrium constant
L	unidentate ligand
LL	bidentate ligand (LL' different donor atoms)
M	central metal atom
nmr	nuclear magnetic resonance
R	alkyl group
uv	ultraviolet
X or Y	unidentate ligand with single negative charge
· *eg* I·	free radical (unpaired electron)
$ML_6, Y^{(n-1)+}$	ion association complex of ML_6^{n+} and Y^-
acacH	pentane-2,4-dione
$bigH_2$	biguanidine
dipy	2,2'-dipyridine
cyclam	1,4,8,11-tetraazacyclotetradecane
$edtaH_4$	bis[di(carboxymethyl)amino] ethane (ethylenediaminetetraacetic acid)
en	ethane-1,2-diamine
phen	1,10-phenanthroline
$P(C_6H_5)_3$	triphenylphosphine
py	pyridine
trien	triethenetetramine

References

1. *Chemical nomenclature, symbols and terminology.* Hatfield: The Association for Science Education, 1972.
2. R. S. Cahn, *An introduction to chemical nomenclature*, 4th edn. London: Butterworths, 1968.

CONTENTS

PREFACE iii

NOMENCLATURE iv

CHAPTER
1. INTRODUCTION 1
Definitions, 1; Terminology, 1; Nucleophilic and electrophilic attack, 1; Free radical reactions, 2; Classification, 2.

2. THE ELUCIDATION OF A MECHANISM 3
The rate equation, 3 ; Confirmation of a mechanism from other than kinetic evidence, 3; detailed knowledge of the nature of the reactants, 3; detection of an intermediate, 4; the stoichiometry and a detailed knowledge of the products, 5.

3. SUBSTITUTION REACTIONS INVOLVING METAL ION COMPLEXES 9
Octahedral complexes, 9; hydrolysis, 12; replacement reactions, 19; reactions in non aqueous solvents, 20; isomerisation and racemisation, 21; Planar four coordinate complexes, 24; Experimental, 26.

4. REDOX REACTIONS INVOLVING METAL IONS 32
Inner and outer sphere reactions, 32; Survey of reaction types, 36; electron exchange reactions, 36; mutual oxidation-reduction reactions, 38; reactions between different metal ions, 38; reactions between metal ions and non metallic species, 42; oxidative addition, 51; Experimental, 52.

5. SOME REACTIONS OF THE P BLOCK ELEMENTS 61
Addition-dissociation reactions, 61; Exchange reactions, 61; Redox reactions, 63; Some reactions of the oxides of nitrogen, 64; dinitrogen pentoxide, 65; nitrogen dioxide, 66; nitrogen oxide, 66; Clock reactions, 67; Experimental, 73.

v

6. METAL ION ASSISTED PROCESSES 76
Reactions at a limited number of sites, 76; Reactions involving stabilisation in a particular configuration, 77; Reactions favoured as a result of internal strain, 77; Reactions involving electron redistribution, 78; Template reactions, 79; Insertion reactions, 80; Oligomerisation and polymerisation, 81; Metal ion catalysed redox reactions, 82; Autoxidation, 83; Experimental, 85.

7. OSCILLATING REACTIONS 90
The Belousov–Zhabotinskii reaction, 92; Experimental, 95.

APPENDIX 1: PREPARATIONS 98

SUGGESTIONS FOR FURTHER READING 99

1. Introduction

Definitions
A reaction mechanism is the detailed stepwise process involving molecules, atoms, radicals or ions that occurs simultaneously or consecutively and culminates in the observed overall reaction. The detail which may be ascertained is dependent on the particular reaction. In the majority of cases it is only possible to elucidate a basic outline for the reaction mechanism. In relatively few cases is it possible to establish a detailed mechanism which is without doubt correct. Inorganic has been taken to cover all reactions involving metal ions and, in addition, all other reactions apart from those containing carbon–hydrogen or carbon–carbon bonds at the reaction site.

Terminology
Many terms well known in organic chemistry are useful in the discussion of inorganic systems. These are summarised briefly.

Nucleophilic and electrophilic attack
Most reagents are polar and consequently the $\delta+$ site will attract nucleophiles, electron rich positive site seeking reagents and the $\delta-$ site electrophiles, electron deficient negative site seeking reagents.

A typical reaction involving nucleophilic attack is represented

$$AB + X^- \to AX + B^-$$

Such a reaction is generally classified as S_N1 (substitution, nucleophilic, unimolecular) or S_N2 (substitution, nucleophilic, bimolecular) depending on the nature of the rate determining step. For example the S_N1 process might be

$$AB \to A^+ + B^- \quad \text{rate determining}$$
$$A^+ + X^- \to AX$$

and the S_N2 process

$$AB + X^- \to ABX^- \quad \text{rate determining}$$
$$ABX^- \to AX + B^-$$

In practice, however, it is not always easy to assign the terms S_N1 or S_N2 to a particular reaction (p 11).

Electrophilic reactions are treated similarly and the classification S_E1 and S_E2 used. These reactions are of less significance in inorganic chemistry where the central atom in a given species is a positive site. The examples encountered generally involve attack on a given ligand (p 19).

Free radical reactions[1]

A free radical is defined as an atom, molecule or complex which contains one or more unpaired electrons. Transition metal ions containing an unpaired electron(s) are excluded.

With the exception of a few stable free radicals such as chlorine dioxide, nitrogen oxide and nitrogen dioxide free radicals in a given reaction arise from homolytic bond fission

$$AB \rightarrow A\cdot + B\cdot$$

Classification

The classification of inorganic reactions is somewhat difficult. Many reactions could equally well be discussed under several headings. The classification adopted in this monograph is, by and large, the generally accepted one, classifying reactions by type: redox, substitution, *etc*, rather than by mechanism: nucleophilic, electrophilic, free radical *etc*.

Reference

1. Organic free radical reactions are discussed by J. I. G. Cadogan in Monograph No. 24, *Principles of free radical chemistry*. London: The Chemical Society, 1973.

2. The Elucidation of a Mechanism

The most important piece of evidence in the elucidation of a reaction mechanism is generally the experimentally determined rate equation. The importance of acquiring all possible additional information cannot, however, be over emphasised. In particular, attention should be paid to:

1. The exact nature (including stereochemistry) of both reactants and products.
2. The presence of any equilibria.
3. The stoichiometry of the reaction.

The rate equation

In this monograph familiarity with the determination and interpretation of a rate law for a given reaction has generally been assumed.[1] It is important, however, to remember that the interpretation is merely a suggestion and is frequently one of several plausible mechanisms. Credibility for a particular suggestion must be obtained from other than kinetic methods.

Confirmation of a mechanism from other than kinetic evidence

Confirmation of a mechanism suggested on the basis of a rate equation may be obtained from:

1. Detailed knowledge of the nature of the reactants.
2. Detection (direct or indirect) of a suspected intermediate.
3. Detailed knowledge of the nature of the products.

These possibilities are illustrated by some examples.

Detailed knowledge of the nature of the reactants
(a) *The copper(I) reduction of iron(III)*

$$Cu(I) + Fe(III) \rightarrow Cu(II) + Fe(II) \qquad 1$$

The behaviour of this reaction is described by the rate law

$$\frac{-d[Fe(III)]}{dt} = \frac{k[Fe(III)][Cu(I)]}{[H^+]}$$

The inverse dependence on $[H^+]$ suggests that a hydrolysis product is the reactant. Knowledge that iron(III) solutions at the given pH contain the hydroxopentaaquairon(III) ion and that hydrolysis under the given conditions is negligible for copper(I) suggests that it is the hydrolysed iron(III) species which is the reactant.

The mechanism is then represented

$$Fe(H_2O)_6^{3+} \rightleftharpoons Fe(H_2O)_5OH^{2+} + H^+ \qquad 2$$
$$Fe(H_2O)_5OH^{2+} + Cu(I) \rightarrow products \qquad 3$$

(b) The formation of urea from ammonium cyanate

$$NH_4^+ + OCN^- \rightarrow OC(NH_2)_2 \qquad 4$$

The behaviour of this reaction is described by the rate law

$$\frac{d[OC(NH_2)_2]}{dt} = k[NH_4^+][OCN^-]$$

A possible interpretation of the rate expression is that the reaction proceeds *via* a simple bimolecular process involving the ammonium and cyanate ions. However, knowledge that ammonium cyanate is in equilibrium with ammonia and isocyanic acid

$$NH_4^+ + OCN^- \rightleftharpoons NH_3 + OCNH \qquad 5$$

suggests that a more likely course for the reaction involves ammonia and isocyanic acid, since ammonia is much more able to attack the appropriate carbon atom than the ammonium ion, which is co-ordinatively saturated. The suggested mechanism is represented

$$\begin{array}{c}H\\H-N:\\H\end{array} + \ddot{O}=C=N-H \longrightarrow \begin{array}{c}H\quad H\\N---H\\|\\\ddot{O}=C---N-H\end{array} \longrightarrow \ddot{O}=C\begin{array}{c}\ddot{N}\diagup^H_H\\\diagdown_{N\diagdown H}^{\ddot{}\;H}\end{array} \qquad 6$$

Detection of an intermediate

In many instances a mechanism postulated from an observed rate equation will involve an intermediate species. Detection and identification of the intermediate will provide strong evidence for a given suggestion. Detection of an intermediate is generally spectroscopic. Ultraviolet–visible spectroscopy and electron spin resonance spectroscopy are particularly useful. Some reactions with intermediates absorbing in the visible region are given in Table 1.

In certain circumstances a particular intermediate may not be readily observable, however, it is sometimes possible to react (trap or scavenge) all (or part) of the intermediate with an added substrate, thereby enabling indirect observation. For successful scavenging of the intermediate the rate of the substrate-intermediate reaction must be greater than or comparable to the other reactions of the intermediate. Furthermore, the substrate must be unreactive towards the reactants and products at least on the

Table 1. Some reactions with visually observable intermediates.

Reactants	Intermediate[a]
Cu(II)—SO_3^{2-}	Green $Cu(SO_3)_n^{2-2n}$ or $Cu(HSO_3)_n^{2-n}$
Cu(I)—$S_2O_3^{2-}$	Yellow-green $Cu(S_2O_3)_{n_1}^{2-2n}$
Fe(III)—$S_2O_3^{2-}$	Brown-violet $Fe(S_2O_3)_{n_2}^{3-2n}$
Fe(III)—cysteine	Blue 1:1 complex coordinated through O and S

[a] Although these species are apparently involved in the respective oxidations it should be noted that observation alone does not imply that the species lies in the reaction pathway.

time scale required for the experiment. The method is not of widespread application but has proved useful in a few instances. The most noteworthy are the detection of tin(III) *via* the reduction of various cobalt(III) complexes and of chromium(V) *via* the oxidation of iodide ion. For example, when the cerium(IV) oxidation of tin(II)

$$2Ce(IV) + Sn(II) \rightarrow 2Ce(III) + Sn(IV) \qquad 7$$

is carried out in the presence of the tris(ethanedionato)cobaltate(III) ion $(Co(C_2O_4)_3^{3-})$ reduction of the cobalt(III) complex is observed. Tin(II) or (IV), cerium(III) or (IV) do not react with the complex in the time scale of interest. Reduction of the cobalt(III) complex is thus taken as evidence for tin(III). Similar behaviour is observed for some other tin(II) reductions which are thus assumed to involve tin(III). It should, however, be noted that failure to detect an intermediate by this method does not provide positive proof of its absence. Failure may merely result from the fact that the rate of the substrate-intermediate reaction is much less than that for some other reaction of the intermediate.

The stoichiometry and a detailed knowledge of the products

Both knowledge of the stoichiometry and details of the exact nature of the products can often provide useful mechanistic information. In a typical investigation the stoichiometry will generally be determined first. This should be examined with each of the reactants in excess. A definite stoichiometry may suggest a single pathway but a variable stoichiometry with variable products will certainly suggest more than one pathway for the reaction. For example, the oxidation of hydrogentrioxosulphate(IV) by various one electron oxidants (*eg* Ce(IV) or Fe(III)) exhibits a stoichiometry HSO_3^- : oxidant of 1.0 to 2.0 depending on the conditions. Both tetraoxosulphate(VI) and dithionate may be obtained as products (p 43).

In certain cases a different but definite stoichiometry is obtained depending upon which reactant is in excess, the products being

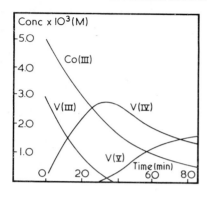

Fig. 1. Dependence of concentrations of reactants and products upon time in the cobalt(III) oxidation of vanadium(III). Initial concentrations $[Co(III)] = 4.99 \times 10^{-3}$ M; $[V(III)] = 2.93 \times 10^{-3}$ M; $[V(IV)] = 3 \times 10^{-5}$ M; $[H^+] = 2.96$ M; $T = 5\,°C$.

different in each case. For example, the oxidation of vanadium(III) by cobalt(III) in the presence of excess vanadium(III) is represented

$$V(III) + Co(III) \rightarrow V(IV) + Co(II) \qquad 8$$

while in the presence of excess cobalt(III) the reaction is represented*

$$V(III) + 2Co(III) \rightarrow V(V) + 2Co(II) \qquad 9$$

Clearly this behaviour is a result of the vanadium(III) reduction of vanadium(V)

$$V(V) + V(III) \rightarrow 2V(IV) \qquad 10$$

The kinetic behaviour is shown in *Fig. 1* where it can be seen that, in the presence of excess vanadium(III), vanadium(V) is removed as fast as it is produced.

Examples illustrating information obtained from routine examination of the stereochemistry (p 14) and chemical constitution (p 33) of the products are provided in more appropriate sections of the text.

By far the most important information obtained from an examination of the products comes from isotopic labelling. The use of a particular isotope enables the observation of bond breaking processes, identification of atom or group transfer and provides a means of probing detailed stereochemical changes. Some examples illustrate these possibilities.

---

* It may be suggested that stoichiometries for reactions of this type can be predicted from a knowledge of the respective electrode potentials; it should, however, be remembered that a reaction may be kinetically controlled and in practice, therefore, there is no substitute for the actual determination.

THE ELUCIDATION OF A MECHANISM

The chlorine oxidation of hydrogen peroxide

$$Cl_2 + H_2O_2 \rightarrow 2H^+ + 2Cl^- + O_2 \qquad 11$$

In this reaction, in aqueous solution, the oxygen may be considered to be derived from the hydrogen peroxide or from the solvent. Using $H_2{}^{18}O_2$ it may be shown that all the oxygen is from the hydrogen peroxide. The suggested mechanism is represented:

$$Cl_2 + H_2{}^{18}O_2 \rightarrow H^+ + Cl^- + H^{18}O^{18}OCl \qquad 12$$

$$H^{18}O^{18}OCl \rightarrow H^+ + Cl^- + {}^{18}O_2 \qquad 13$$

The trioxochlorate(V) oxidation of hydrogentrioxosulphate(IV)

$$ClO_3^- + 3HSO_3^- \rightarrow Cl^- + 3SO_4^{2-} + 3H^+ \qquad 14$$

In this reaction, in aqueous solution, it may be envisaged that the oxygen in the tetraoxosulphate(VI) is derived from the trioxochlorate(V) or the solvent. Using ^{18}O enriched trioxochlorate(V) and normal hydrogentrioxosulphate(IV) it can be seen that the number of oxygen atoms transferred from the trioxochlorate(V) varies from 2.1 to 2.9, depending on the conditions.

The suggested mechanism is represented:

$$HSO_3^- + H^+ \rightleftharpoons H_2SO_3 \qquad 15$$

$$H_2SO_3 \rightleftharpoons SO_2 + H_2O \qquad 16$$

$$ClO_3^- + SO_2 \rightarrow \left[\begin{array}{c} O \\ {}_O\!\!>\!\!Cl\text{---}O\text{---}S\!\!<\!\!{}^O_O \end{array} \right] \rightarrow ClO_2^- + SO_3 \qquad 17$$

$$SO_3 + H_2O \rightarrow H_2SO_4 \qquad 18$$

with subsequent analogous steps.

Two explanations for the transfer of less than three oxygen atoms may be suggested. It is clear that it is the third transfer (involving HClO) which is in doubt. The oxygen exchange (with the solvent) in the case of oxochlorate(I) is rapid and thus it may be that this is responsible. Alternatively, it may be that the third step proceeds, at least in part, *via* the transfer of chlorine rather than oxygen

$$HClO + H^+ + \underset{-O}{\overset{O}{\overset{\|}{S}}}\text{---}OH \rightarrow H_2O + Cl\text{---}\underset{O}{\overset{O}{\overset{\|}{S}}}\text{---}OH \qquad 19$$

$$\xrightarrow{H_2O} 3H^+ + Cl^- + SO_4^{2-}$$

The synthesis and acid hydrolysis of trioxothiosulphate(VI) (thiosulphate)

Using ^{35}S it may be demonstrated that the synthesis of trioxothiosulphate(VI) ion proceeds with retention of the S–O bonds

$$S_8 + 8\,{}^{35}SO_3^{2-} \rightarrow 8S^{35}SO_3^{2-} \qquad 20$$

The reversal of this reaction in acid solution is described by the rate equation

$$\frac{d[SO_2]}{dt} = k[S_2O_3^{2-}][H^+]$$

Using $S^{35}SO_3^{2-}$ all the ^{35}S is found in the sulphur dioxide. These facts are consistent with nucleophilic displacement of trioxosulphate(IV) ion

$$H^+ + S_2O_3^{2-} \rightleftharpoons HS_2O_3^- \qquad 21$$
$$HS^{35}SO_3^- + S^{35}SO_3^{2-} \rightarrow H\text{-}S\text{-}S\text{-}{}^{35}SO_3^- + {}^{35}SO_3^{2-} \qquad 22$$
$$^{35}SO_3^{2-} + 2H^+ \rightleftharpoons H_2{}^{35}SO_3 \rightleftharpoons {}^{35}SO_2 + H_2O \qquad 23$$

successive displacements yielding S_8.

The isomerisation of the nitritopentaamminecobalt(III) ion

$$\underset{\text{pink}}{Co(NH_3)_5ONO^{2+}} \rightleftharpoons \underset{\text{yellow}}{Co(NH_3)_5NO_2^{2+}} \qquad 24$$

This reaction involves a change of the coordinated NO_2^- group from O bonded to N bonded (nitrito isomer to nitro isomer). In aqueous solution using excess (non coordinated) ^{18}O labelled dioxonitrate(III) ion it may be shown that there is no exchange of dioxonitrate(III) during the isomerisation. The mechanism is thus presumed to involve intramolecular displacement

$$(NH_3)_5CoONO^{2+} \longrightarrow (NH_3)_5Co\overset{O}{\underset{N-O}{\diagup\diagdown}}{}^{2+} \longrightarrow (NH_3)_5CoNO_2^{2+} \qquad 25$$

This mechanism has also been suggested for the solid state rearrangement. It should, however, be noted that an intramolecular mechanism is not a prerequisite even for solid state rearrangements. Thus in the solid state the S bonded thiocyanatopentaamminecobalt(III) ion is observed to rearrange to the N bonded isomer with at least 45 per cent of the product formed *via* a dissociative process.

Reference

1. See P. G. Ashmore, *Principles of reaction kinetics*, 2nd edn. London: The Chemical Society, 1973; and J. L. Latham and A. E. Burgess, *Elementary reaction kinetics*. London: Butterworths, 1977.

3. Substitution Reactions Involving Metal Ion Complexes

A metal ion in aqueous solution is surrounded by a number of ligands. The replacement of one ligand by another is an important and extensively studied reaction. Such substitution reactions are frequently of significance in the redox (p 32), isomerisation (p 22) and racemisation (p 22) reactions involving metal ion complexes. This chapter provides a brief survey of the behaviour of six co-ordinate octahedral and four coordinate planar complexes. The commonly encountered tetrahedral complexes undergo rapid ligand exchange reactions and have not as yet received significant attention.

Octahedral complexes

The rate of replacement of the ligands of six coordinate complexes by other ligands varies over a very wide range. As an example, for the exchange of water first order rate constants may range from ca $10^9 \, s^{-1}$ (alkali metals) to ca $10^{-9} \, s^{-1}$ (chromium(III)). The terms labile and inert are used to indicate whether the exchange is rapid or slow but the distinction is not rigid. As a general rule those complexes which react completely within about one minute at 25 °C (for ca 10^{-2} M solutions) are considered labile and those which take longer are considered inert. It should be emphasised that the term inert should only be used in the comparison of rates and is not to be confused with the term stable which refers to thermodynamic stability. Although an inert complex may be stable the two properties are not necessarily related. This point is exemplified by consideration of the inert complex ion hexaamminecobalt(III) where the equilibrium constant for the acid hydrolysis

$$Co(NH_3)_6^{3+} + 6H_3O^+ \rightarrow Co(H_2O)_6^{3+} + 6NH_4^+ \qquad 26$$

is ca 10^{25} but the reaction requires many hours even in six molar hydrochloric acid.

Some generalisations as to the expected rates of substitution reactions are possible.

1. The larger the central atom (other factors being equal) the less tightly held are the ligands and the more labile is the complex.
2. The higher the charge on the central atom (other factors being equal) the more tightly held are the ligands and the more inert is the complex.

The combined effects of charge and size are demonstrated in the series

$$AlF_6^{3-}, SiF_6^{2-}, PF_6^-, SF_6$$

where hydrolysis with dilute sodium hydroxide yields aluminium hydroxide in the case of the hexafluoroaluminate(III) ion but sulphur hexafluoride is resistant to hot concentrated sodium hydroxide. The effect of ligand size is dependent on the type of ligand(s) and on the central metal atom. A particular example is given on p 14.

This section concentrates on transition metal complexes where the degree of lability or inertness of a given complex can be correlated with the electronic configuration (Table 2).

Table 2. Electronic configurations of some octahedral species.

Electronic configuration			Metal ion[a]
nd	$(n+1)$s	$(n+1)$p	
		labile	
(empty)	↑↓ ↑↓	↑↓	↑↓ ↑↓ ↑↓	Ti(IV), Mo(VI), Ce(IV)
1	↑↓ ↑↓	↑↓	↑↓ ↑↓ ↑↓	Ti(III), V(IV)
1	↑↓ ↑↓	↑↓	↑↓ ↑↓ ↑↓	V(III)
		inert	
1 1 1 ↑↓ ↑↓	↑↓	↑↓ ↑↓ ↑↓	V(II), Cr(III)
↑↓ 1 1 ↑↓ ↑↓	↑↓	↑↓ ↑↓ ↑↓	Cr(CN)$_6^{4-}$
↑↓ ↑↓ 1 ↑↓ ↑↓	↑↓	↑↓ ↑↓ ↑↓	Fe(CN)$_6^{3-}$, Fe(phen)$_3^{3+}$, Ru(III)
↑↓ ↑↓ ↑↓ ↑↓ ↑↓	↑↓	↑↓ ↑↓ ↑↓	Co(III), Ru(II), Rh(III), Pd(IV), Pt(IV)

[a] All the examples considered are low spin complexes utilising so far as possible the lower d orbitals. High spin complexes eg Fe(H$_2$O)$_6^{3+}$ and CoF$_6^{3-}$ using $(n+1)$d orbitals are labile.

The principal inert complexes are those of cobalt(III) and chromium(III) and thus, not surprisingly, the majority of studies of substitution reactions involving octahedral complexes have concerned these metal ions. Substitution reactions involving metal complexes are conveniently classified in the categories: hydrolysis, replacement reactions and reactions in non-aqueous solvents. These will be discussed, principally for reactions involving unidentate ligands, under these headings. Firstly, for cobalt(III) where the majority of information is available and then to a lesser extent for other metal ions. Initially, however, it is advantageous to examine a general case represented by the reaction

$$L_5MX^{n+} + Y^- \rightarrow L_5MY^{n+} + X^- \qquad 27$$

where L is a non-involved ligand.
For this process two limiting cases may be considered (Table 3).

Table 3. Possible mechanisms for the substitution reaction 27.

Dissociative	Associative
$L_5M\ X^{n+} \rightarrow L_5M^{(n+1)+} + X^-$ five coordinate intermediate	$L_5M\ X^{n+} + Y^- \rightarrow L_5M\ XY^{(n-1)+}$ seven coordinate intermediate
$L_5M^{(n+1)+} + Y^- \rightarrow L_5M\ Y^{n+}$	$L_5MXY^{(n-1)+} \rightarrow L_5MX^{n+} + X^-$
Classification	Classification
S_N1 (substitution, nucleophilic unimolecular)	S_N2 (substitution, nucleophilic, bimolecular)

It should be stressed that these mechanisms are extreme cases and that reactions frequently occur by a mechanism which lies between the two extremes. This possibility has been termed a concerted or interchange(I) mechanism in which it is considered that the incoming ligand occupies an appropriate place in the solvation shell of the complex, then substitutes at the metal centre without having made any real contribution to the activation process. Cases of this type may be considered where the activation energy is affected by the leaving group (Id-dissociative process) or where the activation energy is affected by the entering group (Ia-associative process). This discussion has, however, been largely restricted to the terms S_N1 and S_N2 and uses these terms to indicate not extreme cases but reactions which approximate to the stated behaviour.

The expected rate laws for the two cases are

1. Dissociative: $\quad -\dfrac{d[L_5MX^{n+}]}{dt} = k[L_5MX^{n+}]$

2. Associative: $\quad -\dfrac{d[L_5MX^{n+}]}{dt} = k[L_5MX^{n+}][Y^-]$

Unfortunately, however, a particular rate law does not prove that the reaction is S_N1 or S_N2. For example, solvent intervention yields the mechanism

$$L_5MX^{n+} + H_2O \rightarrow L_5MH_2O^{(n+1)+} + X^- \qquad \text{slow} \qquad 28$$

$$L_5MH_2O^{(n+1)+} + Y^- \rightarrow L_5MY^{n+} + H_2O \qquad \text{fast} \qquad 29$$

and the rate law
$$-\frac{d[L_5MX^{n+}]}{dt} = k[L_5MX^{n+}]$$

(The concentration of water (~ 55.5 M) is effectively invariant during the reaction.) The reaction, however, could be S_N1 or S_N2.

A further possibility arises from formation of ion association complexes leading to a pre-equilibrium of the type

$$L_5MX^{n+} + Y^- \rightleftharpoons L_5MX,Y^{(n-1)+} \qquad 30$$

where the reactant species are associated although Y has not entered the coordination sphere of M. The subsequent decomposition

$$L_5MX,Y^{(n-1)+} \rightarrow ML_5Y^{n+} + X^- \qquad 31$$

would yield a second order rate equation although the reaction is unimolecular.

In the case of base hydrolysis yet another possibility is significant (p 16).

Hydrolysis

Hydrolysis is generally classified according to the conditions. In acid solution the process is termed acid hydrolysis or aquation and is illustrated by the equation

$$L_5MX^{n+} + H_2O \rightarrow L_5MH_2O^{(n+1)+} + X^- \qquad 28$$

In basic solution the process is termed base hydrolysis and is illustrated by the equation

$$L_5MX^{n+} + OH^- \rightarrow L_5MOH^{n+} + X^- \qquad 32$$

Depending on the pH of the reaction mixture it follows that the product of a given hydrolysis can be a mixture of both the aqua and hydroxo complexes.

For a typical complex for which base hydrolysis is observable the rate law obeyed is

$$-\frac{d[L_5MX^{n+}]}{dt} = k_A[L_5MX^{n+}] + k_B[L_5MX^{n+}][OH^-]$$

The first term (k_A) refers to the acid hydrolysis and the second term (k_B) to the base hydrolysis. These processes are conveniently discussed separately.

Acid hydrolysis

For hydrolysis at pH 0–3 the base hydrolysis term is negligible (k_B is typically 10^5–10^8 times k_A) and the behaviour is first order

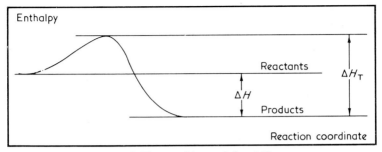

FIG. 2. The transition enthalpy ΔH_T.

with respect to the complex and independent of the acid concentration.* The rate law provides no information as to the role of water and does not enable a distinction between a dissociative, associative or concerted process to be obtained. Mechanistic information must, therefore, be obtained from other than kinetic sources. Some approaches to the elucidation of the mechanistic details of these reactions have been extremely ingenious but, unfortunately, almost every interpretation may be countered with a plausible alternative. It is, however, apparent that in the majority of cases, for cobalt(III), the process is a dissociative one. Some of the evidence is presented as follows:

1. The rates of hydrolysis correlate fairly well (inversely) with the thermodynamic bond strength of the Co–X bond indicating that this bond is broken initially.
2. The transition enthalpy ΔH_T, defined as the enthalpy difference between the transition state and products (*Fig. 2*), is independent of the nature of the leaving group for the complexes $Co(NH_3)_5X^{2+}$ where $X = Cl^-$, Br^- and NO_3^-. This is explicable if the mechanism is dissociative involving a common intermediate.
3. The rate of substitution for the process

$$Co(LL)_2Cl_2^+ + H_2O \rightarrow Co(LL)_2(H_2O)Cl^{2+} + Cl^- \qquad 33$$

   increases with an increase in size of the ligand LL (Table 4). This is explicable if the process is dissociative when an increase in size of the non involved ligands would discourage association to give a seven coordinate intermediate.
4. In certain cases it is possible to confirm that an intermediate in the hydrolysis (presumed $Co(NH_3)_5^{3+}$) is capable of discriminating between various ligands in solution. For example, the

---

* In certain cases where the ligand concerned has retained some basicity the rate may be acid dependent.

Table 4. Rate constants for the acid hydrolysis of trans Co(LL)$_2$-Cl$_2^+$ at 25 °C.

LL	$k \times 10^5$ (s^{-1})
NH$_2$CH$_2$CH$_2$NH$_2$	3.2
NH$_2$CH$_2$CH(CH$_3$)NH$_2$	6.2
dlNH$_2$CH(CH$_3$)CH(CH$_3$)CH(CH$_3$)NH$_2$	15
meso NH$_2$CH(CH$_3$)CH(CH$_3$)NH$_2$	42
NH$_2$CH$_2$C(CH$_3$)$_2$CH$_2$NH$_2$	300

azidopentaamminecobalt(III) ion reacting with dioxonitric(III) acid in the presence of a second anion X$^-$ (F$^-$, Cl$^-$, Br$^-$, NO$_3^-$ or SCN$^-$) yields Co(NH$_3$)$_5$X^{2+} in addition to the aquapentaamminecobalt(III) ion. The suggested mechanism for the acid (HNO$_2$) hydrolysis of azidopentaamminecobalt(III) (Co(NH$_3$)$_5$-N$_3^{2+}$) in the presence of anion X$^-$ is

$$Co(NH_3)_5N_3^{2+} + HNO_2 \longrightarrow Co(NH_3)_5N_3NO^{3+}$$

$$\downarrow$$

$$Co(NH_3)_5^{3+} + N_2O + N_2 \qquad 34$$

$$\swarrow_{H_2O} \qquad \searrow^{X^-}$$

$$Co(NH_3)_5(H_2O)^{3+} \qquad Co(NH_3)_5X^{2+}$$

5. Aquation is frequently accompanied by stereochemical change (Table 5). It is suggested that this is a consequence of the geometry of the transition state. The dissociative mechanism proceeding *via* a five coordinate intermediate will result in stereochemical change if the intermediate is a trigonal bipyramid but not if the structure is square pyramidal (Scheme 1). The fact that some complexes show a stereochemical change on

Table 5. Steric course of the aquation of some cobalt(III) complexes at 25 °C.

L$_5$CoX^{n+} + H$_2$O → L$_5$Co(H$_2$O)$^{(n+1)+}$ + X$^-$		28
L$_5$CoX^{n+}	% Cis	% Trans
trans Co(NH$_3$)$_4$Cl$_2^+$	55	45
trans Co(en)$_2$OHCl$^+$	75	25
trans Co(en)$_2$Cl$_2^+$	35	65
trans Co(en)$_2$NCS Br$^+$	45	55
trans Co(cyclam)OHCl$^+$	0	100
trans Co(cyclam)Cl$_2^+$	0	100
cis Co(en)$_2$Cl$_2^+$	100	0
cis Co(en)$_2$OHCl$^+$	100	0
cis Co(en)$_2$NH$_3$Cl^{2+}	84	16

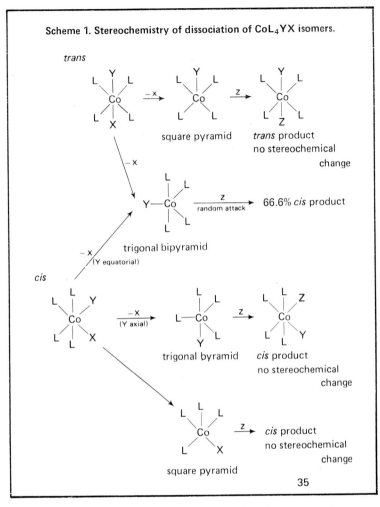

Scheme 1. Stereochemistry of dissociation of CoL_4YX isomers.

aquation while others do not is considered to result from the delicate balance which exists between the two structures.

It is evident that there is a significant effect in the case of ligands which have an unshared pair of electrons in addition to the pair used in formation of the coordinate bond (eg OH^-, Cl^-, NCS^-). Cis complexes with a non-replaced ligand of this type show no stereochemical rearrangement. This is explained by suggesting that the unshared pair of electrons is available for donation to the metal in a p–d pi bond and that such pi bonding stabilises the square pyramidal intermediate by lowering the positive charge on the metal. When the same ligand is in the *trans* position orbital overlap is not possible unless the complex rearranges to a trigonal bipyramidal structure, the consequence of which is stereochemical change. *Trans* complexes which show no rearrangement are presumed to hydrolyse *via* a square pyramidal intermediate.

Base hydrolysis

The dependence of the rate of hydrolysis on the concentration of hydroxide ion (p 12) indicates an associative (S_N2) mechanism. However, although this accounts for the observed kinetics it does not indicate why:

1. Hydroxide ion is unique in its effect on the rate of replacement of a given ligand, other nucleophiles do not show such an increase in rate.
2. Complexes in which none of the ligands have ionisable hydrogen atoms *eg* $Co(py)_4Cl_2^+$ and $Co(CN)_5Cl^{3-}$ are generally hydrolysed only slowly and at a rate independent of the hydroxide ion concentration.

To explain these observations a mechanism in which the hydroxide ion promotes the reaction through proton abstraction, and which proceeds *via* a five coordinate intermediate, has been suggested. For the chloropentaamminecobalt(III) ion the mechanism is represented

$$Co(NH_3)_5Cl^{2+} + OH^- \underset{}{\overset{fast}{\rightleftharpoons}} (NH_3)_4Co(NH_2)Cl^+ + H_2O \qquad 36$$

$$(NH_3)_4Co(NH_2)Cl^+ \overset{slow}{\longrightarrow} (NH_3)_4Co(NH_2)^{2+} + Cl^- \qquad 37$$

$$(NH_3)_4Co(NH_2)^{2+} + H_2O \overset{fast}{\longrightarrow} (NH_3)_5CoOH^{2+} \qquad 38$$

This mechanism is classified as S_N1CB (substitution, nucleophilic, unimolecular, conjugate base) and, at least in the case of cobalt(III) complexes, considerable evidence in favour of this behaviour has been accumulated.

The fact that direct attack by hydroxide ion does not occur has been demonstrated through a study of the behaviour of naturally occurring ^{18}O (present as *ca* 0.2 per cent oxygen) in the hydrolysis of complex ions of the formula $Co(NH_3)_5X^{2+}$. It is observed that ^{18}O has a slight preference for water molecules rather than hydroxide ions. Hydroxide ions in aqueous solution thus have a smaller abundance of ^{18}O than the surrounding water molecules. This effect enables the source of oxygen, from hydroxide ion or water, to be obtained as, in aqueous solution, we have

$$\frac{[H_2{}^{18}O][{}^{16}OH^-]}{[H_2{}^{16}O][{}^{18}OH^-]} = 1.040$$

Measurement of this fractionation factor for the product of base hydrolysis of the pentaamminecobalt(III) ion yields

$$\frac{[Co(NH_3)_5{}^{16}OH^{2+}][H_2{}^{18}O]}{[Co(NH_3)_5{}^{18}OH^{2+}][H_2{}^{16}O]} = 1.0056$$

If hydroxide ion had entered the coordination sphere directly the value would have been 1.040 while direct attack by water (as required by the five coordinate intermediate in the conjugate base mechanism) would have yielded a value of 1.000. Further evidence for the presence of a five coordinate intermediate is analogous to that for the aquation process. The hydrolysis is often accompanied by stereochemical change. Steric crowding of the metal ion centre generally increases rather than decreases the rate as expected for an associative process. In suitable circumstances it is possible to observe discrimination of the intermediate between various ligands. For example, when the nitratopentaamminecobalt(III) ion $(Co(NH_3)_5NO_3^{2+})$ is treated with sodium hydroxide in the presence of appreciable thiocyanate the S bonded thiocyanatopentaamminecobalt(III) ion $(Co(NH_3)_5SCN^{2+})$ is found in the products. The presence of coordinated thiocyanate is indicative of a five coordinate intermediate while the S bonded product is indicative of a preference of the intermediate for sulphur rather than nitrogen. The more normally encountered product is the N bonded isothiocyanatopentaamminecobalt(III) ion to which the S bonded species isomerises. The behaviour is summarised in Scheme 2. Observation of products other than $Co(NH_3)_5OH^{2+}$

Scheme 2. Suggested mechanism for the base hydrolysis of the nitratopentamminecobalt(III) ion in the presence of thiocyanate ion.

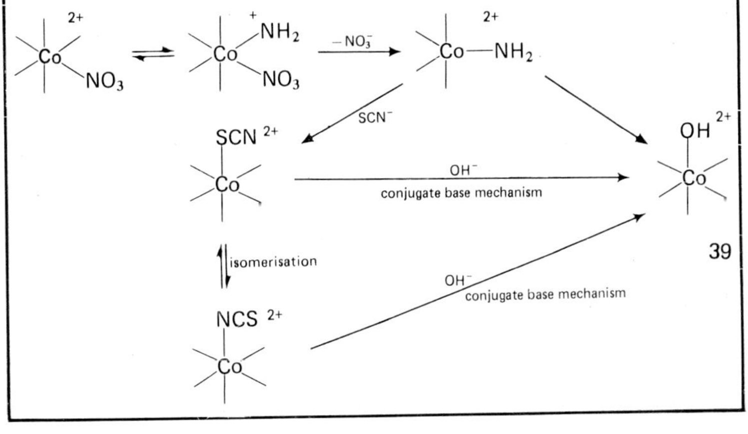

is possible because the base hydrolysis of the SCN and NCS coordinated species is substantially slower than that of $Co(NH_3)_5NO_3^{2+}$. The products may thus be trapped by acidification after, say, 4–6 half lives for the base hydrolysis of $Co(NH_3)_5NO_3^{2+}$.

Direct evidence for the conjugate base has proved difficult to obtain in view of the extremely low concentration present. Recent work has, however, enabled some information to be obtained and a pK_a value for the tris(ethane-1,2-diamine)cobalt(III) ion of *ca* 14.2 has been obtained from nmr measurements. The low concentration of conjugate base indicates its very high reactivity towards water. The reasons for this reactivity are not altogether clear but may be suggested as

1. Due to the lower charge on the conjugate base, thus increasing the lability of the coordinated negative ion.
2. Due to the stabilisation of the five coordinate intermediate through a pi bonding contribution:

$$\begin{array}{c}H\\ \diagdown\\ N\!\!\rightarrow\!\!X\\ H\diagup\;\;\;|\\ Co\end{array}\!\!\!{}^{+}\quad\longrightarrow\quad \begin{array}{c}H_2N\diagdown\\ Co\end{array}\!\!\!{}^{2+}\;+\;X^- \qquad\qquad 40$$

General discussion

Although the overwhelming majority of evidence for hydrolysis in the case of cobalt(III) complexes favours a dissociative process for acid hydrolysis and the S_N1CB dissociative process for the base hydrolysis, it is emphasised that the situation is far from completely settled and there are certainly exceptions. The cobalt(III) complex $Co(edta)^-$ is, for example hydrolysed at a rate proportional to the hydroxide ion concentration although there are no ionisable hydrogen atoms in this complex and the behaviour thus cannot be S_N1CB. An associative process involving seven coordinate $CoedtaOH^{2-}$ is suggested. Seven coordinate complex ions of this type have been successfully isolated for a number of first row transition metal ions though apparently not for cobalt(III).

As yet a further possibility, in a few instances it has been demonstrated that the aquation process does not involve the cleavage of metal-ligand bonds, but instead bonds within a ligand are broken and reformed. This behaviour is observed in the aquation of the carbonatopentaamminecobalt(III) ion which, using ^{18}O labelled water, may be shown to proceed

$$(NH_3)_5Co-OCO_2 + 2H_3{}^{18}O^+ \rightarrow (NH_3)_5CoH_2O^{3+} + 2H_2{}^{18}O + CO_2 \qquad 41$$

The reaction apparently involves electrophilic attack on the oxygen atom bound to the cobalt followed by subsequent loss of carbon dioxide and protonation of the hydroxo complex (42).

$$(NH_3)_5Co-O-C\overset{O^+}{\underset{O}{\diagup}} \longrightarrow (NH_3)_5CoOH^{2+} + CO_2$$
$$\underset{H^+}{|}$$
$$\overset{O}{\underset{H\;\;\;\;H}{\diagup}}$$
$$\downarrow H_3O^+$$
$$Co(NH_3)_5H_2O^{3+}$$

42

Metal ions other than cobalt(III)

The acid hydrolysis of complexes of chromium(III), nickel(II), copper(II), ruthenium(II), ruthenium(III) and rhodium(III) has been examined in some detail. The majority of cases appear to involve a dissociative mechanism. Some aquations, for example those of $Cr(LL)_2Cl_2^+$, where the rate is effectively independent of LL (compare cobalt(III) Table 4) do, however, appear associative. Details of the base hydrolysis of complexes other than those of cobalt(III) are not clear. The S_N1CB mechanism has been suggested for numerous chromium(III) complexes, there are, however, exceptions and certainly the effect is less pronounced than in the case of cobalt(III).

Replacement reactions

Reactions considered under this heading are:

1. Replacement of water by a given ligand, a process often termed anation.
2. Replacement of a ligand other than water.

Reactions of both types are frequently described by the term complex formation. Elucidation of the mechanistic details is often complicated by the formation of ion association complexes (p 12) but this may, however, be overcome by examination of anionic complexes such as $Co(CN)_5(H_2O)^{2-}$.

For the anation

$$Co(CN)_5H_2O^{2-} + X^- \rightarrow Co(CN)_5X^{3-} + H_2O \qquad 43$$

the rate is observed to be proportional to $[X^-]$ at low concentration of X^- but is effectively independent of $[X^-]$ at high concentration. At high concentration of X^- the rate is identical to that for exchange of water in the complex. The mechanism is thus considered dissociative

$$Co(CN)_5H_2O^{2-} \rightleftharpoons Co(CN)_5^{2-} + H_2O$$
$$\downarrow X^-$$
$$Co(CN)_5X^{3-}$$

(44)

It appears that the majority of replacement reactions involving cobalt(III) are analogous and involve a five coordinate intermediate. However, it has been suggested that some reactions of ruthenium(III), rhodium(III) and iridium(III) are associative. For example, the reaction

$$Ru(NH_3)_6^{3+} + NO + H^+ \rightarrow Ru(NH_3)_5NO^{3+} + NH_4^+$$

(45)

occurs much more rapidly than could be explained by aquation of the hexaammineruthenium(III) ion.

Bidentate and multidentate ligands

Bidentate and multidentate ligands are generally bound more strongly than unidentate ones (chelate effect). Consequently, a bidentate or multidentate ligand dissociates considerably more slowly than a unidentate one. Dissociation of such ligands requires rotation of the chelate ring to allow a water (solvent) molecule to enter the coordination sphere and prevent the chelate bond reforming. A number of systems have been investigated, principally those involving the ligands 2,2′-dipyridine, ethane-1,2-diamine, ethanedioate and edta^{4-} and the metal ions Cr(III), Fe(II), Fe(III), Co(III) and Ni(II). These reactions unlike those involving unidentate ligands are acid catalysed. The catalysis apparently involves protonation of the first donor atom dissociated. A typical reaction is represented

$$\begin{array}{c}\text{[M(en)]}^{n+} \rightleftharpoons \text{[M(NH_2CH_2CH_2NH_2)]}^{n+} \rightarrow \text{[M]}^{n+} + NH_2CH_2CH_2NH_2 \\ \downarrow H^+ \\ \text{[M(NH_2CH_2CH_2NH_3)]}^{(n+1)+} \xrightarrow{fast} \text{[M]}^{n+} + NH_2CH_2CH_2NH_3^+ \end{array}$$

(46)

Reactions in non aqueous solvents

Although the majority of work on substitution reactions has been carried out in aqueous solution some information is available for solvents such as methanol and dimethylsulphoxide. Interest in the use of non-aqueous solvents arises from the possibility of examining the role of water in the substitution process, as in replacement processes of the type

$$Co(en)_2Cl_2^+ + X^- \rightarrow Co(en)_2XCl^+ + Cl^- \qquad 47$$

In aqueous solution these reactions proceed *via* the aquated complex

$$Co(en)_2Cl_2^+ \xrightarrow[H_2O]{-Cl^-} Co(en)_2(H_2O)Cl^{2+} \xrightarrow[-H_2O]{X^-} Co(en)_2XCl^+ \qquad 48$$

The mechanism is apparently dissociative in both steps. In methanol, a poor coordinating solvent, no formation of an intermediate methanol complex is observed and strong evidence for an S_N1 dissociative process is obtained as substitution by trioxonitrate(v), bromide, and thiocyanate all proceed at the same rate. Unfortunately, however, these observations would also be satisfied if the process was S_N2 involving a labile, highly unstable, species containing coordinated methanol

$$Co(en)_2Cl_2^+ + CH_3OH \rightleftharpoons Co(en)_2CH_3OHCl^{2+} + Cl^-$$
$$\downarrow X^- \qquad 49$$
$$Co(en)_2XCl^+ + CH_3OH$$

Significant evidence against this latter process has been obtained by the preparation of $Co(en)_2CH_3OHCl^{2+}$ which, although labile, is not sufficiently reactive to account for its absence in the replacement reactions. The mechanism is thus considered to involve an S_N1 dissociative process.

Isomerisation and racemisation

The most widely studied reactions are those of $Co(en)_2X_2^+$, $Co(en)_2XY^+$ and tris chelate complexes of the type $M(LL)_3$ and $M(LL')_3$. It is apparent that these processes generally involve a dissociative mechanism. (Examples of linkage isomerisation have been given on p 8.)

Cis-trans isomerisation

Typical examples of this type of reaction are the isomerisation of the *cis*-dichlorobis(ethane-1,2-diamine)cobalt(III) ion and the *trans*-hydroxoaminebis(ethane-1,2-diamine)cobalt(III) ion. The isomerisation of the *cis*-dichlorobis(ethane-1,2-diamine)cobalt(III) ion is

$$Co(en)_2Cl^+ \rightleftharpoons Co(en)_2Cl_2^+ \qquad 50$$
$$\text{cis} \qquad \text{trans}$$
$$\text{(purple)} \qquad \text{(green)}$$

Isotopic tracer experiments indicate that exchange of chloride ion accompanies this isomerisation and provide evidence that the exchange is not a direct chloride ion for chloride ion. The isomerisation is thought to proceed *via* the equilibria

$$Co(en)_2Cl_2^+ + H_2O \rightleftharpoons Co(en)_2(H_2O)Cl^{2+} + Cl^- \qquad 51$$

$$Co(en)_2(H_2O)Cl^{2+} + H_2O \rightleftharpoons Co(en)_2(H_2O)_2^{3+} + Cl^- \qquad 52$$

Apparently, opening of the chelate ring does not occur. The significant intermediate seems to be the five coordinate trigonal bypyramidal species $Co(en)_2Cl^{2+}$

$$\text{(structure)} \xrightleftharpoons{-Cl^-} \text{(structure)} \xrightleftharpoons{Cl^-} \text{(structure)} \qquad 53$$

The isomerisation of the trans-*hydroxoamminebis(ethane-1,2-diamine)-cobalt(III) ion* $(Co(en)_2NH_3OH^{2+})$

In this instance, the isomerisation occurs without exchange of ammonia or hydroxide ion. It is evident that complete dissociation of either ethane-1,2-diamine group does not occur as this would lead to decomposition of the complex not isomerisation. The methanism is thus intramolecular and the pathway assumed to involve partial dissociation of ethane-1,2-diamine to yield a unidentate ligand. The proposed mechanism is represented

$$\text{(structure)} \rightleftharpoons \text{(structure)} \rightleftharpoons \text{(structure)} \qquad 54$$

Racemisation

An example of a racemisation reaction is provided by the tris-(ethanedionato)chromium(III) ion

$$\text{(structure)} \rightleftharpoons \text{(structure)} \qquad 55$$

In this instance, the racemisation is faster than the ligand exchange reaction. Interestingly, all 12 oxygen atoms in the complex exchange with solvent water at a much faster rate than that for ethanedionate exchange. Clearly, this indicates a ring opening process. The oxygen exchange is acid catalysed and the proposed mechanism is

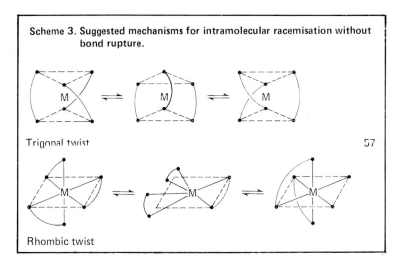

56

Although most examples of racemisation involve dissociation this is not the only possibility. Intramolecular processes without bond rupture have been suggested for racemisation of Fe(phen)$_3^{2+}$, Co(bigH)$_3^{3+}$ and Co(edta)$^-$. In such cases a so-called twist mechanism is suggested. Two possibilities (Scheme 3) the trigonal or Bailar twist and the rhombic or Ray-Dutt twist have received particular attention.

Scheme 3. Suggested mechanisms for intramolecular racemisation without bond rupture.

Trigonal twist

57

Rhombic twist

Planar four coordinate complexes

Most information on substitution reactions of this geometry is concerned with platinum(II). The other systems that have been investigated are Rh(I), Ir(I), Pd(II), Ni(II), Cu(II)* and Au(III). For a typical substitution

$$L_3PtX^+ + Y^- \rightleftharpoons L_3PtY^+ + X^- \qquad 58$$

the rate law observed is

$$\frac{-d[L_3PtX^+]}{dt} = \{k_1 + k_2[Y^-]\}[L_3PtX^+]$$

The two pathways are considered separately.

The k_1 pathway

The rate in this pathway is independent of the concentration of the entering ligand and, as for the octahedral case, the rate determining step may be considered as either associative or dissociative (Table 6).

Several factors suggest that the mechanism is associative. For example,

1. Steric congestion generally retards the k_1 (and k_2) pathway.
2. The k_1 term is very sensitive to the nature of the solvent, indicative of bond making rather than bond breaking and, of course, indicative of solvent intervention.

The k_2 pathway

It is generally agreed that this pathway involves an S_N2 (associative) process. It is difficult to observe the five coordinate *intermediate* and it has been reported only in a few instances, for reactions involving rhodium(I) and nickel(II). Salts containing five coordinate species are, however, relatively common and examples involving platinum(II), nickel(II) and rhodium(I) have been reported. Both square pyramidal and trigonal bipyramidal structures have been noted and, in at least one instance ($Ni(CN)_5^{3-}$), the energy

---

* The majority of commonly encountered copper(II) complexes are distorted octahedral (*eg* $Cu(NH_3)_4(H_2O)_2^{2+}$) or distorted tetrahedral (*eg* $CuCl_4^{2-}$ in Cs_2CuCl_4).

Table 6. Associative and dissociative mechanisms for the substitution reaction 58.

Dissociative	Associative
$L_3PtX^+ \rightarrow L_3Pt^{2+} + X^-$ three coordinate intermediate	$L_3PtX^+ + Y^- \rightarrow L_3PtXY$ five coordinate intermediate
$L_3Pt^{2+} + Y^- \rightarrow L_3PtY^+$	$L_3PtXY \rightarrow L_3PtY^+ + X^-$

Associative with solvent intervention

$L_3PtX^+ + S \rightarrow L_3PtXS^+$
  five coordinate
  intermediate

$L_3PtXS^+ \rightarrow L_3PtS^{2+} + X^-$

$L_3PtS^{2+} + Y^- \rightarrow L_3PtSY^+$
  five coordinate
  intermediate[a]

$L_3PtSY^+ \rightarrow L_3PtY^+ + S$

[a] A dissociative step at this stage could also be considered.

difference between the two forms is small enough to permit isolation of a stable salt, ($[Cr(en)_3][Ni(CN)_5]1.5H_2O$), the crystals of which have two $Ni(CN)_5^{3-}$ ions per unit cell, one square pyramidal and one trigonal bipyramidal.

For both pathways substitution proceeds with retention of configuration. This is anticipated for formation of a five coordinate intermediate in the substitution $ML_2X_2 + Y^- \rightarrow ML_2XY + X^-$

59

The interesting observation that the *cis–trans* isomerism in planar complexes is catalysed by traces of free ligands is, therefore, explained in terms of a two-stage mechanism. For example, in the case of dichlorobis(triethylphosphine)platinum(II) the mechanism

is

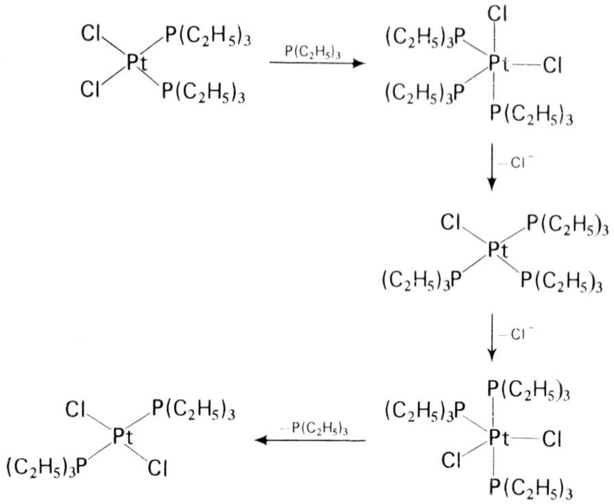

60

Experimental

The hydrolysis of some chromium(III) and cobalt(III) complexes

These reactions are conveniently followed colorimetrically or conductimetrically. In some cases sampling and volumetric analysis (visual or potentiometric end point detection) is possible. A large number of complexes are suitable for examination. A particularly convenient compound on account of its availability is dichlorotetraaquachromium(III) chloride-2-water.*

The hydrolysis of the dichlorotetraaquachromium(III) ion

This complex is subject to both acid and base hydrolysis. The observed rate expression is

$$\frac{-d[Cr(H_2O)_4Cl_2^+]}{dt} = k_A[Cr(H_2O)_4Cl_2^+] + k_B[Cr(H_2O)_4Cl_2^+][OH^-]$$

The aquation is represented

$$Cr(H_2O)_4Cl_2^+ + H_2O \rightarrow Cr(H_2O)_5Cl^{2+} + Cl^- \quad 61$$

$$Cr(H_2O)_5Cl^{2+} + H_2O \rightarrow Cr(H_2O)_6^{3+} + Cl^- \quad 62$$

In the presence of appreciable acid, replacement of the second chloride is not significant in the time scale required for observation of the first stage. Under these conditions the reaction is conveniently investigated *via* sampling and potentiometric analysis for chloride ion.

* This is commercial chromium(III) chloride–$CrCl_3.6H_2O$. The configuration is *trans*.

Kinetic investigation (30–40 °C)

Requirements

[Cr(H$_2$O)$_4$Cl$_2$]Cl.2H$_2$O*

HClO$_4$ 2.5×10^{-1} M

AgNO$_3$ 10^{-2} M (1.70 g dm^{-3})

A potentiometer with galvanometer, a voltmeter, or pH meter calibrated in millivolts; silver and calomel electrodes.

Prepare a 10^{-2} M (2.66 g dm^{-3}) solution of dichlorotetraaquachromium(III)chloride-2-water in 2.5×10^{-1} M tetraoxochloric(VII) acid at the required temperature (30–40 °C). Titrate potentiometrically 10.0 cm^3 aliquots *vs* time (*eg* every 10–15 min) with 10^{-2} M silver nitrate. The aliquots may be quenched in cold 2.5×10^{-1} M tetraoxochloric(VII) acid if required.

A value of the infinity titre (V_∞) for the first aquation to [Cr(H$_2$O)$_5$Cl^{2+}] may be calculated from the mass of [Cr(H$_2$O)$_4$Cl$_2$]Cl.2H$_2$O used. At the pH examined the base hydrolysis term is negligible. A plot of log $V_\infty - V_t$ *vs* time enables the first order rate constant (k_A) to be obtained (-2.303.gradient).

Literature value $k_{25°C}$ 8.3×10^{-5} s^{-1}.

Typically $k_{36.5°C}$ 20–30×10^{-5} s^{-1}.

Table 7. Complexes suitable for investigation.

Complex	Method	1st order rate constant (s^{-1}) 25 °C	Ref.
Co(NH$_3$)$_5$Cl^{2+}	As above, and colorimetric	1.7×10^{-6}	1
Co(C$_2$O$_4$)$_3^{3-}$	Colorimetric	6.2×10^{-5}	2

1. R. J. Angelici, *Synthesis and technique in inorganic chemistry*. Philadelphia: Saunders, 1969.
2. P. W. Wiggans, *Educ. Chem.*, 1975, **12**, 54.

Some other complexes suitable for investigation in an analogous manner are shown in Table 7.

In the absence of added acid the hydrolysis of complexes of this type is conveniently followed conductimetrically. In practice both acid and base hydrolysis is observed under these conditions. In the case of the dichlorotetraaquachromium(III) ion, treatment as acid hydrolysis yields a convenient if not strictly correct illustration of the method. The first order rate constant thus obtained can be used, with the literature value for K_A and the pH of the solution, to calculate K_B.

*This is commercial chromium(III) chloride–CrCl$_3$.6H$_2$O. The configuration is *trans*.

Kinetic investigation (25 °C)

Requirements

[Cr(H$_2$O)$_4$Cl$_2$]Cl.2H$_2$O
Conductance bridge and cell.
Record the conductivity *vs* time for a 10^{-3}M (0.266 g dm^{-3}) solution of dichlorotetraaquachromium(III) chloride-2-water at 25 °C.
Over the first 6–7 min the second stage of the aquation can be ignored. An infinity value for the first aquation may be obtained from the molar conductivity (10^{-3} M) of [Cr(H$_2$O)$_5$Cl]Cl$_2$.2H$_2$O which is 266 Ohm^{-1} cm^2 mol^{-1} at 25 °C. A plot of log $C_\infty - C_t$ *vs* time yields (-2.303 gradient) a first order rate constant. Typically $k_{25°C}$ 1–7 × 10^{-3} s^{-1}.

The chromium(III)-edta H$_2^{2-}$ reaction

This reaction is generally represented

$$\text{Cr(H}_2\text{O)}_6^{3+} + \text{edtaH}_2^{2-} \rightarrow \text{Cr(edta)}^- + 4\text{H}_2\text{O} + 2\text{H}_3\text{O}^+ \qquad 63$$

In practice, however, the product involves pentacoordinate edtaH^{3-} and the sixth coordination position is occupied by a water molecule

The reaction is thus actually represented

$$\text{Cr(H}_2\text{O)}_6^{3+} + \text{edtaH}_2^{2-} \rightarrow \text{HCr(edta)H}_2\text{O} + 4\text{H}_2\text{O} + \text{H}_3\text{O}^+ \qquad 64$$

Kinetic investigation (20–30 °C).

Requirements

Na$_2$edtaH$_2$
NaOH 10^{-1} M (4.00 g dm^{-3})
KCr(SO$_4$)$_2$.12H$_2$O 10^{-2} M (49.9 g dm^{-3})
Colorimeter: Filter 495–575 nm (Ilford No. 624) or Spectrophotometer $\lambda = 545$ nm.

Prepare a solution 10^{-1} M (37.22 g dm^{-3}) with respect to the disodium salt of edtaH$_2$ and at a pH of 5.0 (adjusted with sodium hydroxide). To 10 cm^3 of this solution add 10 cm^3 10^{-2} hexaaquachromium(III). Record the absorbance *vs* time curve for the

resultant mixture. Readings are conveniently taken every 5 min (25 °C). An 'infinity' value may be obtained by boiling the solution. A plot of log $A_\infty - A_t$ vs time yields (-2.303.gradient) the first order rate constant. Typically $k_{25°C}$ 1–4 × 10^{-4} s^{-1}.
An ethanoate (acetate) buffer may be used but there is some spectrophotometric evidence for involvement of the ethanoate ion. The effect of pH (3.5–5.5) and chromium(III) concentration may be examined. The rate expression observed is

$$\frac{d[\text{complex}]}{dt} = k[\text{Cr(III)}][\text{H}^+]^{-1}$$

Above pH6 replacement of the coordinated water by hydroxide ion yields blue Cr(edta)OH^{2-}. Potentiometric investigation[1] of this behaviour provides a useful introduction to a kinetic investigation.

The presence of a free carboxyl group in the chromium(III) complex may be demonstrated kinetically by a comparison of the rate of oxidation of the ligand by the tetraoxomanganate(VII) ion in the presence and absence of bismuth(III).

The rate of the tetraoxomanganate(VII) oxidation of edtaH$_{4-n}^{n-}$ is reduced by coordination of the edtaH$_{4-n}^{n-}$. Addition of bismuth(III) to the chromium(III) complex further decreases the rate of oxidation. Clearly the bismuth(III) ties up some important part of the ligand which is still available in the chromium(III) complex (a ligand exchange reaction is not feasible in view of the inert nature of the chromium(III)). It is assumed that the free carboxyl group in the chromium(III) complex is coordinated to the bismuth(III) and is thus less available for attack by the tetraoxomanganate(VII) ion. The rate determining step in the tetraoxomanganate(VII) oxidation of edtaH$_{4-n}^{n-}$ apparently involves attack at a free carboxyl group.

Kinetic investigation (20–25 °C)

Requirements

KMnO$_4$ 2 × 10^{-2} M (3.16 g dm^{-3})
HCr (edta)H$_2$O 10^{-2} M* (4.39 g dm^{-3})
H$_2$SO$_4$ 1M
Bi(NO$_3$)$_3$.5H$_2$O 10^{-2} M in 1 M H$_2$SO$_4$ (4.85 g dm^{-3} 1M H$_2$SO$_4$)
Colorimeter: Filter 495–575 nm (Ilford No. 624) or Spectrophotometer $\lambda = 545$ nm.

---

* This solution may be obtained directly from the solid salt[1] or prepared by boiling a solution containing the appropriate equimolar quantities of chromium-(III) and the disodium salt of edtaH$_4$.

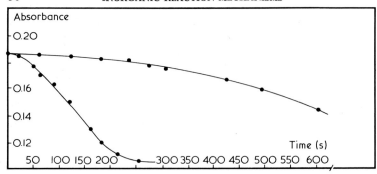

FIG. 3. Absorbance vs time curves for the tetraoxomanganate(VII) oxidation of the chromium(III) complex Cr(edta)H$_2$O$^-$. (Lower curve) in the absence of ismuth(III). (Upper curve) in the presence of bismuth(III). Conditions as in text. $T = 20\,°C$.

(*i*) Oxidation in the absence of bismuth(III)
Record the absorbance vs time curve for the reaction initiated by addition of 0.2 cm^3 2×10^{-2} M potassium tetraoxomanganate(VII) to a solution of 10 cm^3 10^{-2} M HCr(edta)H$_2$O and 10 cm^3 1 M tetraoxosulphuric(VI) acid made up to 100 cm^3 with distilled water.

(*ii*) Oxidation in the presence of bismuth(III)
Repeat (*i*) but replace the 10 cm^3 1 M tetraoxosulphuric(VI) acid by a solution 10^{-2} M with respect to bismuth(III) and 1 M with respect to tetraoxosulphuric(VI) acid.
Typical results are shown in *Fig. 3*.

Some reactions of the trans-dichlorobis(ethane-1,2-diamine)cobalt(*III*) *ion*
Requirements
NaNO$_2$ 1 M (69.0 g dm^{-3})
NH$_3$ (aq) 2.5×10^{-2} M
[Co(en)$_2$Cl$_2$]Cl* 10^{-2} M (2.85 g dm^{-3}). This solution must be freshly prepared.

Demonstration
Replacement of chloride ion by dioxonitrate(III) ion or ammonia is suitable for demonstration.

(A)   trans Co(en)$_2$Cl$_2^+$ + NO$_2^-$ → Co(en)$_2$ClNO$_2^+$ + Cl$^-$ 65
                green   yellow

Mix equal volumes of 1 M sodium dioxonitrate(III) and 10^{-2} M *trans*-dichlorobis(ethane-1,2-diamine)cobalt(III) chloride.

---

* For preparation see Appendix 1.

The exchange is complete in *ca* 1 min.

(B)   *trans* $Co(en)_2Cl_2^+$ + NH_3 → $Co(en)_2NH_3Cl^{2+}$ + Cl^- 66
                     green   red

Mix equal volumes of 2.5×10^{-2} M ammonia and 10^{-2} M *trans*-dichlorobis(ethane-1,2-diamine)cobalt(III) chloride.

The exchange is complete in *ca* 1 min.

Reference
1. J. I. Hoppé and P. J. Howell, *Educ. Chem.*, 1975, **12**, 12.

4. Redox Reactions Involving Metal Ions

Inner and outer sphere reactions

Redox reactions involving metal ions are customarily discussed in terms of electron transfer and written as half cell equations

$$M^{n+} \rightleftharpoons M^{(n+1)+} + e^- \qquad \qquad 67$$

there is, however, no evidence that in aqueous solutions the electron from a reducing agent is released and becomes solvated before reacting with an oxidising agent. The reactions are generally classified as outer sphere or inner sphere reactions. In the outer sphere case both metal ions retain their inner coordination sphere. In the inner sphere case the two metal ions are linked by a common bridging ligand at the time of electron transfer. It is clear that the outer sphere process although not necessarily observed is open to all redox reactions. Distinction between the two mechanisms is, however, not always possible.

Demonstration of an outer sphere mechanism requires:

1. Observation of a rate law first order in both species, corresponding to an activated complex containing all the ligands in the coordination spheres of both ions.
2. Observation that the rate of electron transfer is faster than the rate of substitution into the coordination sphere of either of the metal ions.

The inner sphere mechanism is most easily demonstrated in cases where either the oxidant or reductant is substitutionally inert and after electron transfer the initially labile partner is inert to substitution. For example, the chromium(II) reduction of cobalt(III) complexes

$$\begin{array}{cccc} Cr(II) + & Co(III) \rightarrow & Cr(III) + & Co(II) \\ \text{labile} & \text{inert} & \text{inert} & \text{labile} \end{array} \qquad 68$$

For reduction of the chloropentaamminecobalt(III) ion the rate equation observed is

$$\frac{-d[Co(NH_3)_5Cl^{2+}]}{dt} = k[Co(NH_3)_5Cl^{2+}][Cr(H_2O)_6^{2+}]$$

corresponding to an activated complex involving one mole of each species. Examination of the products shows that the chromium(III) is present as the chloropentaaquachromium(III) ion

$$Co(NH_3)_5Cl^{2+} + Cr(H_2O)_6^{2+} + 5H_3O^+$$
$$\rightarrow Co(H_2O)_6^{2+} + Cr(H_2O)_5Cl^{2+} + 5NH_4^+ \qquad 69$$

Since chromium(III) is inert to substitution it is clear that the chlorine–chromium bond must be formed at, or prior to, the electron transfer. Furthermore, it may be demonstrated, using radioactive chloride ion, that all the chlorine in the chromium product is from the cobalt(III) complex. The activated complex is represented

$$\left[\begin{array}{c}NH_3\\H_3N\diagdown\,|\,\diagup NH_3\\Co\\H_3N\diagup\,|\,\diagdown\\NH_3\end{array}H_2O\begin{array}{c}OH_2\\\diagdown\,|\,\diagup OH_2\\Cr\\\diagup\,|\,\diagdown OH_2\\OH_2\end{array}\right]^{4+}$$

with a Cl bridge between Co and Cr.

In certain instances where both of the products are inert to substitution a bridged dimer may actually be isolated as the product. Thus, crystalline salts of the ion $[(CN)_5CoNCFe(CN)_5]^{6-}$ may be isolated from the hexacyanoferrate(III) oxidation of the pentacyanocobaltate(II) ion.

Additional complexities are introduced when the bridging ligand affords more than one site for coordination. For example, in the chromium(II) reduction of the thiocyanatopentaamminecobalt(III) ion ca 30 per cent of the product is the thiocyanatopentaaquachromium(III) ion and the remainder the isothiocyanatopentaaquachromium(III) ion. The suggested mechanism involves bridge formation at both the nitrogen of the thiocyanate group (remote attack) and at the sulphur (adjacent attack)

$$(NH_3)_5CoSCN^{2+} + Cr(II) \longrightarrow \begin{array}{c}[(NH_3)_5Co-S(-N\equiv C)-Cr(H_2O)_4]^{4+} \longrightarrow (H_2O)_5CrSCN^{2+}\\ \\ [(NH_3)_5Co-SCN-Cr(H_2O)_4]^{4+} \longrightarrow (H_2O)_5CrNCS^{2+}\end{array}$$

70

A distinction may be drawn between inner sphere reactions where the rate determining step is either

(a) substitution ie bridge formation

or

(b) electron transfer within the bridged intermediate.

Both cases generally follow simple second order kinetics; the difference may, however, be inferred from the rate constants. For case (a) the rate of electron transfer will be comparable to the rate of substitution (provided of course that the outer sphere mechanism is not favoured). It is also anticipated that this mechanism will not be particularly sensitive to the nature of the oxidant (or reductant as appropriate). For case (b) where one component is substitutionally labile the bridge formation may be considered as rapid and reversible. The subsequent relatively slow rearrangement is then expected to be sensitive to the oxidant (or reductant as appropriate). It is clear that satisfactory conclusions cannot always be made. Some examples where a reasonable distinction is possible are given in Table 8.

Table 8. Second order rate constants for some oxidations at 25 °C.

Reductant	Oxidant	$k(dm^3\ mol^{-1}\ s^{-1})$
V(II)	$Fe(H_2O)_5Cl^{2+}$	4.6×10^5
V(II)	$Ru(H_2O)_5Cl^{2+}$	1.9×10^3
V(II)	$Ru(NH_3)_5Cl^{2+}$	3.0×10^3
V(II)	Cu(II)	26.6
V(II)	$Co(NH_3)_5Cl^{2+}$	7.6
V(II)	$Co(CN)_5N_3^{3-}$	101.0
Cr(II)	$Co(NH_3)_5CH_3COO^{2+}$	1.8×10^{-1}
Cr(II)	$Co(NH_3)_5Cl^{2+}$	6.0×10^5
Cr(II)	$Co(NH_3)_5CN^{2+}$	3.6×10^1
Cr(II)	$Co(NH_3)_5I^{2+}$	3.4×10^6

The examples are presented in three groups. For group 1 the observed rates for the redox reactions are very much greater than the rate of aquation of vanadium(II) ($k_{25°C}$ 100 s^{-1}). These reactions must proceed *via* an outer sphere mechanism. For the second group the rate is comparable to the rate of aquation and thus these reactions are considered to be substitution limited. For the third group the reductant chromium(II) is labile and the reactions are assumed to involve rapid and reversible formation of the bridged intermediate followed by a slow electron transfer. The large variation in rate with the oxidant is to be noted.

It can be supposed that the inner sphere mechanism proceeds *via* atom (or group) transfer but this is not the only possibility. For example, in the chromium(II) reduction of the *cis*-bis(azidotetraaqua)chromium(II) ion two azido bridges are formed although only one electron is transferred.

$$Cr(H_2O)_6^{2+} + {}^*Cr(H_2O)_4(N_3)_2^+ \longrightarrow \begin{bmatrix} Cr \overset{N=N=N}{\underset{N=N=N}{\Large\diagup\!\!\!\diagdown}} Cr \end{bmatrix}^{3+}$$

$$\downarrow$$

$${}^*Cr(H_2O)_6^{2+} + Cr(H_2O)_4(N_3)_2^+$$

71

Furthermore, in the case of the platinum(II)–platinum(IV) exchange in chloride media only one bridge is formed but two electrons are transferred. In several instances, for example

$$Co^{II}(edta)^{2-} + Fe(CN)_6^{3-} \rightarrow [(edta)CoNCFe(CN)_5]^{5-} \quad 72$$
$$\rightarrow Co^{III}(edta)^{-} + Fe^{II}(CN)_6^{4-}$$

it may be demonstrated that bridge formation occurs but the bridging ligand is not transferred.

Returning to the outer sphere case it is of interest to consider the factors affecting electron transfer by this process. Again, perhaps not surprisingly, bridging ligands are often implicated. For example, the behaviour of the chromium(II) reduction of the inert complex hexaammine cobalt(III) chloride is described by

$$\frac{-d[Co(NH_3)_6^{3+}]}{dt} = [Cr(II)][Co(NH_3)_6^{3+}]\{k_1 + k_2[Cl^-]\}$$

In this instance the cobalt(III) complex does not possess a suitable ligand for bridge formation (*coordinated ammonia* has no lone pair of electrons available). The rate, although *ca* 10^{10} times less than that for the corresponding reduction of the chloropentaamminecobalt(III) ion, is still indicative of an outer sphere mechanism in view of the extreme inert nature of the cobalt(III) complex. The chloropentaaquachromium(III) ion is detected in the products and the chloride catalysed pathway is thus presumed to involve an ion association complex with the activated complex presumed to be $Co(NH_3)_6, ClCr(H_2O)_5^{5+}$. Further examples of this behaviour are considered on p 37. Where bridging does not occur the two inner coordination spheres of the reactants may be considered as approaching within the Van der Waals radii during the formation of the activated complex. Under these conditions the rate of electron transfer is greater for complexes containing ligands such as 1,10-phenanthroline and cyanide than for those containing ligands such as water or ammonia. Hence, a ligand that can delocalise the 'nominally' metal electron(s) significantly lowers the magnitude of the barrier to electron transfer.

Finally, in a discussion of inner and outer sphere mechanisms, it should be emphasised that, in many cases, more than one pathway is observed. Thus, in the chromium(II) reduction of the bis-(pentane-2,4-dionato) ethane-1,2-diaminecobalt(III) ion, both singly and doubly bridged intermediates have been formed as well as the products of an outer sphere process. The suggested mechanism

is

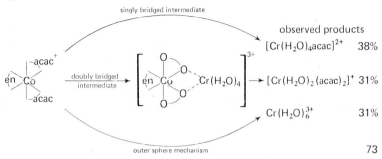

observed products
$[Cr(H_2O)_4acac]^{2+}$ 38%
$[Cr(H_2O)_2(acac)_2]^+$ 31%
$Cr(H_2O)_6^{3+}$ 31%

73

Survey of reaction types

Redox reactions involving metal ions are conveniently discussed as follows:

1. Electron exchange reactions.
2. Mutual oxidation–reduction reactions.
3. Reactions between different metal ions.
4. Reactions between metal ions and non metallic species.

Electron exchange reactions

A typical example is the iron(II)–iron(III) exchange

$$*Fe(II) + Fe(III) \rightleftharpoons *Fe(III) + Fe(II)$$ 74

The behaviour in chloride media is described by

$$\frac{d[Fe(II)]}{dt} = [*Fe(II)][Fe(III)]\left(k_1 + \frac{k_2}{[H^+]} + k_3[Cl^-]\right)$$

The three terms (rate constants k_1, k_2 and k_3) correspond to three different pathways. The first term (k_1) is assumed to result from a direct ion–ion transfer of the outer sphere type. The second and third terms are considered to result from inner sphere pathways involving hydroxide and chloride ions as bridging ligands†

$$*Fe(II) + FeOH^{2+} \rightarrow *Fe(III) + Fe(II) + OH^-$$ 75

$$*Fe(II) + FeCl^{2+} \rightarrow *Fe(III) + Fe(II) + Cl^-$$ 76

---

† The difference in substitutional lability between the two oxidation states in this and other instances provides an explanation for catalysed substitution reactions of the type

$$Fe(H_2O)_5Cl^{2+} + H_2O \xrightarrow{Fe(II)} Fe(H_2O)_6^{3+} + Cl^-$$ 78

$$Cr(H_2O)_5Cl^{2+} + H_2O \xrightarrow{Cr(II)} Cr(H_2O)_6^{3+} + Cl^-$$ 79

As discussed previously reactions where the electron transfer is faster than is feasible for inner sphere bridge formation must proceed *via* an outer sphere mechanism. For example, the exchange between cobalt(II) and cobalt(III) hexammines

$$*Co(NH_3)_6^{3+} + Co(NH_3)_6^{2+} \rightarrow *Co(NH_3)_6^{2+} + Co(NH_3)_6^{3+} \qquad 77$$

The observed rate equation is

$$-\frac{d[*Co(NH_3)_6^{3+}]}{dt} = \frac{k[*Co(NH_3)_6^{3+}][Co(NH_3)_6^{2+}]}{[H^+]}$$

The inverse dependence on hydrogen ion concentration indicates that hydroxide ions are involved in bridging to the inert cobalt(III) complex, presumably through formation of the ion association complex $Co(NH_3)_6, OH^{2+}$ whose uv spectra has been found.

Ion association complexes involving cations and *anionic* complexes have also been proposed to explain experimental observations. For example, the exchange reaction

$$*MnO^- + MnO_4^{2-} \rightleftharpoons *MnO_4^{2-} + MnO^- \qquad 80$$

is known to proceed *via* an outer sphere mechanism but is appreciably influenced by group 1A cations, the acceleration in rate being in the order $Cs > K > Na > Li$. For caesium the behaviour is described by

$$\frac{-d[*MnO_4^-]}{dt} = [*MnO_4^-][MnO_4^{2-}]\{k_1 + k_2[Cs^+]\}$$

indicative of two pathways, one involving a direct ion–ion transfer and the other involving mediation by caesium ions. The precise role of cations in reactions of this type is not clear but has generally been assumed to be that of reduction of coulombic repulsion between the reactants. There is no evidence that the cations actually participate in the electron transfer.

The examples discussed so far have involved direct transfer between the two reactants, it is possible although rare that a more complex mechanism involving disproportionation may provide a competing pathway. For example, the behaviour of the silver(I)–silver(II) exchange

$$*Ag(I) + Ag(II) \rightleftharpoons *Ag(II) + Ag(I) \qquad 81$$

is described by

$$\frac{-d[*Ag(I)]}{dt} = k[Ag(II)]^2$$

It is thus clear that silver(I) is not involved in the rate determining step. The mechanism suggested implicates silver(III) formed in the rate determining step

$$2Ag(II) \rightleftharpoons Ag(I) + Ag(III) \qquad 82$$

$$Ag(III) + Ag(I) \rightarrow 2Ag(II) \qquad 83$$

Exchange reactions involving two or three electrons have not received a great deal of attention. They are often very slow or effectively non-existent (at least at room temperature) eg lead(IV)⇌ lead(II) or chromium(VI)⇌chromium(III), or proceed *via* a dimer in equilibrium with the two species, eg the tin(IV)–tin(II) exchange in chloride media.

$$SnCl_6^{2-} + SnCl_4^{2-} \rightleftharpoons Sn_2Cl_{10}^{4-} \qquad 84$$

Mutual oxidation–reduction reactions

These reactions typified by

$$Mn(VII) + 4Mn(II) \rightarrow 5Mn(III) \qquad 85$$

$$Cr(VI) + 3Cr(II) \rightarrow 4Cr(III) \qquad 86$$

$$Np(V) + Np(III) \rightleftharpoons 2Np(IV) \qquad 87$$

are subject to the same considerations as those reactions between different metal ions and thus have not been treated separately.

Reactions between different metal ions

Reactions in which the oxidant and reductant change oxidation state by the same number of electron equivalents are termed complementary. When the oxidant and reductant differ in their oxidation state changes the reactions are termed non-complementary. Some examples are given below.

Complementary reactions

$$Cu(I) + Fe(III) \rightarrow Cu(II) + Fe(II) \qquad 1$$

$$Cr(II) + Fe(III) \rightarrow Cr(III) + Fe(II) \qquad 88$$

$$Mn(IV) + Tl(I) \rightarrow Tl(III) + Mn(II) \qquad 89$$

Non-complementary reactions

$$Cr(VI) + 3Fe(II) \rightarrow 3Fe(III) + Cr(III) \qquad 90$$

$$2Fe(III) + Sn(II) \rightarrow Sn(IV) + 2Fe(II) \qquad 91$$

$$Mn(VII) + 5Fe(II) \rightarrow 5Fe(III) + Mn(II) \qquad 92$$

Complementary reactions

Complementary reactions involving a one electron change are generally straightforward and are illustrated by the copper(I) reduction of iron(III) (p 3). A somewhat more complex example is afforded by the vanadium(III) reduction of iron(III)

$$Fe(III) + V(III) \rightarrow Fe(II) + V(IV) \qquad 93$$

In the presence of excess iron(II) the rate law observed is

$$\frac{-d[Fe(III)]}{dt} = k[Fe(III)][V(III)] + \frac{k'[Fe(III)][V(III)][V(IV)]}{[Fe(II)]}$$

The second term is not observed in the absence of an excess of one of the products as the concentration of vanadium(IV) is then always equal to that of iron(II) and the expression reduces to the expected second order equation

$$\frac{-d[Fe(III)]}{dt} = k''[Fe(III)][V(III)] \qquad (k'' = k + k')$$

This interesting behaviour is interpreted as resulting from the mechanism

$$Fe(III) + V(III) \xrightarrow{k_1} Fe(II) + V(IV) \qquad 94$$

$$Fe(III) + V(IV) \underset{k_{-2}}{\overset{k_2}{\rightleftharpoons}} Fe(II) + V(V) \quad (\text{note } k_{-2} \gg k_2) \qquad 95$$

$$V(V) + V(III) \xrightarrow{k_3} 2V(IV) \qquad 10$$

which on application of the steady state $(d[V(V)]/dt = 0)$ theory yields the rate equation

$$\frac{-d[Fe(III)]}{dt} = \frac{k_2 k_3 [Fe(III)][V(III)][V(IV)]}{k_{-2}[Fe(II)] + k_3[V(III)]}$$

which is in agreement with the observed second term given that

$$k_{-2}[Fe(II)] \gg k_3[V(III)]$$

For a complementary reaction, such as the uranium(IV) reduction of thallium(III), it is possible to consider the reaction as proceeding *via* a single two electron step or via two one electron steps.

$$U(IV) + Tl(III) \rightarrow U(V) + Tl(II) \qquad 96$$

$$U(V) + Tl(II) \rightarrow U(VI) + Tl(I) \qquad 97$$

Observation of one or both intermediates either directly or *via* trapping experiments (p 4) will enable confirmation of consecutive one electron steps. In the absence of such observations the reaction is regarded as proceeding *via* a single two electron transfer.

Occasionally, kinetic evidence is available to confirm a two electron step, for example, the molecular oxygen oxidation of vanadium(II)

$$O_2 + 2V(II) + 4H^+ \rightarrow 2V(IV) + 2H_2O \qquad 98$$

is very much faster than the corresponding oxidation of vanadium-(III)

$$O_2 + 4V(III) + 4H^+ \rightarrow 4V(IV) + 2H_2O \qquad 99$$

The oxidation of vanadium(II) thus does not involve vanadium(III) and is, therefore, considered to involve a single two electron step.

Non-complementary reactions

Considered as involving only bimolecular steps these reactions must involve an intermediate oxidation state. The possibilities are considered by examining some specific cases.

The iron(II) reduction of thallium(III)

$$2Fe(II) + Tl(III) \rightarrow Tl(I) + 2Fe(III) \qquad 100$$

Two possible schemes are shown in Table 9.

Table 9. Possible mechanisms for the iron(II)–thallium(III) reaction.

Reaction scheme	Rate equation
	$\dfrac{d[Tl(II)]}{dt} = 0 \quad \dfrac{d[Fe(IV)]}{dt} = 0$
	$-\dfrac{d[Tl(III)]}{dt}$
Initial one electron step **A.** (i) $Tl(III) + Fe(II) \underset{k_{-1}}{\overset{k_1}{\rightleftharpoons}} Tl(II) + Fe(III)$ (ii) $Tl(II) + Fe(II) \overset{k_2}{\rightarrow} Tl(I) + Fe(III)$	$\dfrac{k_1 k_2 [Tl(III)][Fe(II)]^2}{k_{-1}[Fe(III)] + k_2[Fe(II)]}$
Initial two electron step **B.** (i) $Tl(III) + Fe(II) \underset{k_{-1}}{\overset{k_1}{\rightleftharpoons}} Tl(I) + Fe(IV)$ (ii) $Fe(II) + Fe(IV) \overset{k_2}{\rightarrow} 2Fe(III)$	$\dfrac{k_1 k_2 [Tl(III)][Fe(II)]^2}{k_{-1}[Tl(I)] + k_2[Fe(II)]}$

It is clear that distinction between the two schemes on the basis of kinetic evidence relies on the reversible nature of A(i) or B(i). The observed behaviour is described by

$$\frac{-dFe[(II)]}{dt} = \frac{k[Fe(II)]^2[Tl(III)]}{[Fe(II)] + k'[Fe(III)]}$$

The fact that the rate is suppressed by iron(III) but that thallium(I) has no effect indicates that the reactive intermediate is thallium(II) and that the mechanism is A.

Interestingly, the analogous chromium(II) reduction

$$2Cr(II) + Tl(III) \rightarrow 2Cr(III) + Tl(I) \qquad 101$$

exhibits simple second order behaviour and no retardation by products is observed. The chromium(III) product $[(H_2O)_4Cr(OH)_2-Cr(H_2O)_4]^{4+}$, however, suggests the presence of chromium(IV). The behaviour is thus apparently analogous to scheme B with 102

$$Cr(II) + Tl(III) \rightarrow Tl(I) + Cr(IV) \qquad 102$$

rate determining and effectively irreversible.

The dimercury(I) reduction of thallium(III)

$$Hg_2^{2+} + Tl(III) \rightarrow 2Hg(II) + Tl(I) \qquad 103$$

The observed behaviour is described by

$$\frac{-d[Tl(III)]}{dt} = \frac{k[Tl(III)][Hg_2^{2+}]}{[Hg(II)]}$$

Separate experiments show that no reaction is observed between thallium(I) and mercury(II). An indirect reaction must thus be proposed to explain the retarding effect of mercury(II). The suggested mechanism involves disproportionation of the dimercury(I)

$$Hg_2^{2+} \rightleftharpoons Hg(II) + Hg(0) \qquad 104$$

$$Hg(0) + Tl(III) \rightarrow Hg(II) + Tl(I) \qquad \text{rate determining} \qquad 105$$

The iron(II) reduction of chromium(VI)

$$Cr(VI) + 3Fe(II) \rightarrow Cr(III) + 3Fe(III) \qquad 90$$

This reaction requires the transfer of three electrons and is thought to proceed *via* three one electron steps

$$Cr(VI) + Fe(II) \rightleftharpoons Cr(V) + Fe(III) \qquad \text{rapid equilibrium} \qquad 106$$

$$Cr(V) + Fe(II) \rightarrow Cr(IV) + Fe(III) \qquad \text{rate determining} \qquad 107$$

$$Cr(IV) + Fe(II) \rightarrow Cr(III) + Fe(III) \qquad 108$$

At low chromium(VI) concentration the rate equation observed is

$$\frac{-d[HCrO_4^-]}{dt} = \frac{kk'[H^+]^3[Fe(II)]^2[HCrO_4^-]}{k''[Fe(III)] + k'[Fe(III)][H^+]}$$

At high chromium(VI) concentration a term in $[Cr(VI)]^2$ presumably resulting from a pathway involving the μ-oxo-bistrioxochromate(VI) ion (dichromate ion) is observed.

Despite the obvious complexity, this reaction is interesting in that considerable information can be obtained from indirect evidence. This is illustrated by examination of the reaction in the presence of iodide ion. When iron(II) is added to a weakly acidic solution containing iodide ion and chromium(VI) iodine is rapidly produced. Under these conditions the direct reaction between chromium(VI) and iodide ion is slow. Two explanations for the rapid production of iodine may be suggested.

(A) $\quad\quad\quad Cr(VI) + 3Fe(II) \rightarrow Cr(III) + 3Fe(III) \quad\quad\quad$ 90

$\quad\quad\quad\quad\quad 2Fe(III) + 2I^- \rightarrow 2Fe(II) + I_2 \quad\quad\quad\quad\quad$ 109

This would imply that iron(II) is a catalyst and would yield a stoichiometry identical to that of the uncatalysed chromium(VI) oxidation of iodide ion.

$$2Cr(VI) + 6I^- \rightarrow 2Cr(III) + 3I_2 \quad\quad\quad 110$$

(B) An intermediate formed in the chromium(VI)–iron(II) reaction is responsible for oxidation of the iodide ion

$$Cr(VI) \xrightarrow[\text{fast}]{Fe(II)} \text{reactive species} \xrightarrow[\text{fast}]{I^-} I_2 \quad\quad 111$$

Such a mechanism indicates that iron(II) is not a catalyst but is consumed stoichiometrically. Reactions of this type are termed induced reactions. A detailed examination of the system shows that the iron(III) oxidation of iodide ion is substantially slower than the production of iodine *via* the chromium(VI)–iron(II)–iodide reaction. Scheme A may, therefore, be discarded. A plot of mole ratio $[I^-]_r:[Fe(II)]_r$ *vs* mole ratio $[I^-]_0:[Fe(II)]_0$, where $_0$ and $_r$ refer to initial and reacted concentrations respectively shows that the stoichiometry $Fe(II):I^-$ approaches 1:2 (*Fig. 4*). The limiting reaction observed is thus

$$Cr(VI) + Fe(II) + 2I^- \rightarrow Cr(III) + I_2 + Fe(III) \quad\quad 112$$

The mole ratio $Cr(VI):Fe(II)$ of 1:1 suggests that a one electron transfer is initially involved

$$Cr(VI) + Fe(II) \rightarrow Cr(V) + Fe(III) \quad\quad 106$$

Reactions between metal ions and non metallic species

The non metallic species which can be examined are numerous and may range from simple halide ions to large organic molecules. Generally, these reactions are more complex than reactions between two metal ions. The non metallic species are usually two electron oxidants or reductants and one electron steps will consequently lead to free radicals and the possibility of additional products

REDOX REACTIONS INVOLVING METAL IONS

through, eg, dimerisation of the radicals or their reaction with oxygen. Such reactions involving one electron steps are hence often characterised by a variable stoichiometry. Some possibilities are illustrated below.

The oxidation of hydrogen trioxosulphate(IV)

This oxidation may proceed *via* two pathways depending on whether the initial step is a one electron or two electron transfer.

$$HSO_3^- \xrightarrow{-2e^-} SO_3 \xrightarrow{H_2O} SO_4^{2-}$$
$$HSO_3^- \xrightarrow{-e^-} HSO_3^\cdot \xrightarrow{HSO_3^\cdot} S_2O_6^{2-}$$
(113)

For transition metal ions such as Cr(VI), Mn(III), Mn(VII), Fe(III) and Co(III) one electron steps are observed and dithionate and tetraoxosulphate(VI) can usually be detected as products. The stoichiometry is somewhat dependent on the actual conditions employed.

Considering the iron(III) oxidation as a comparatively simple example the reaction is represented as a combination of

$$HSO_3^- + 2Fe(III) + H_2O \rightarrow 2Fe(II) + 3H^+ + SO_4^{2-} \quad (114)$$

$$2HSO_3^- + 2Fe(III) \rightarrow 2Fe(II) + 2H^+ + S_2O_6^{2-} \quad (115)$$

In the presence of a large excess of iron(III) the stoichiometry corresponds to (114) and the mechanism

$$H_2SO_3 \rightleftharpoons H^+ + HSO_3^- \quad (116)$$
$$Fe(III) + HSO_3^- \rightleftharpoons Fe(HSO_3)^{2+} \quad (117)$$
$$Fe(HSO_3)^{2+} \rightarrow Fe(II) + HSO_3^\cdot \quad (118)$$
$$HSO_3^\cdot + Fe(II) \rightarrow Fe(III) + HSO_3^- \quad (119)$$
$$HSO_3^\cdot + Fe(III) + H_2O \rightarrow SO_4^{2-} + Fe(II) + 3H^+ \quad (120)$$

is suggested.

In the absence of a large excess of iron(III) removal of $HSO_3^\cdot$ radicals *via* 120 is in competition with 121.

$$2HSO_3^\cdot \rightarrow 2H^+ + S_2O_6^{2-} \quad (121)$$

and dithionate thus appears in the products.

Two electron oxidants do not yield dithionate.

The oxidation of hydrazine

The behaviour of hydrazine as a reductant, in common with hydrogen trioxosulphate(IV), is characterised by whether or not the initial step involves a one or two electron transfer.

For the one electron case we have

$$N_2H_4 \xrightarrow{-e^-} N_2H_3^{\cdot} + H^+ \qquad 122$$

$$2N_2H_3^{\cdot} \rightarrow N_4H_6 \qquad 123$$

$$N_4H_6 \rightarrow N_2 + 2NH_3 \qquad 124$$

while for the two electron case we have

$$N_2H_4 \xrightarrow{-2e^-} N_2H_2 + 2H^+ \qquad 125$$

$$2N_2H_2 \rightarrow N_4H_4 \qquad 126$$

$$N_4H_4 \rightarrow N_2 + N_2H_4 \qquad 127$$

Obviously, further complications arise if successive oxidations occur in preference to dimerisation of the radical species. Not surprisingly the one electron reductions are generally characterised by a variable stoichiometry while the two electron reactions are not (Table 10). In both cases it is evident from the acid dependence that the reactive species are protonated. Spectroscopic observation ($\lambda = 230$ nm) of an intermediate presumed $N_2H_3^{\cdot}$ ($N_2H_4^{\cdot+}$) has been reported for the one electron oxidation while di-imide (N_2H_2) has been detected by mass spectrometry and scavenging experiments, in the two electron oxidations.

Table 10. Stoichiometries observed for the oxidation of hydrazine at pH 0–1.

Oxidising agent	M : N_2H_4	Electrons transferred from the metal ion
Ce(IV)	1.0–1.5 : 1	1.0–1.5
Co(III)	1.0–1.2 : 1	1.0–1.2
Fe(III)	1.0–1.8 : 1	1.0–1.8
Tl(III)	2.0 : 1	4.0
Mn(VII)	0.28–0.44 : 1	1.4–2.2
Cr(VI)	1.1–1.3 : 1	3.5–4.0

The oxidation of chloride, bromide and iodide ion

These metal ion oxidations are often more complex than might be expected. Many possible systems have, perhaps surprisingly, received little or no attention. Again both one and two electron steps are possible. For the one electron process we have

$$X^- \xrightarrow{-e^-} X^{\cdot} \xrightarrow{X^-} X_2^{\cdot -} \qquad 128$$

$$X_2^{\cdot -} \xrightarrow{-e^-} X_2 \qquad 129$$

while for the two electron process we have

$$X^- + H_2O \xrightarrow{-2e^-} HXO + H^+ \qquad 130$$

$$H^+ + HXO + X^- \rightleftharpoons H_2O + X_2 \qquad 131$$

Those reactions where a one electron transfer is involved include the cerium(IV) oxidation of chloride, bromide and iodide ion, the manganese(III) oxidation of bromide and iodide ion and the iron(III) oxidation of iodide ion. Two electron oxidations are apparently operative in the vanadium(V), chromium(VI) and managanese(VII) reactions. Some possibilities are illustrated by discussion of the iron(III), vanadium(V) and bismuth(V) oxidation of iodide ion.

The iron (III) oxidation of iodide ion

$$2Fe(III) + 2I^- \rightarrow 2Fe(II) + I_2 \qquad 109$$

The rate law observed is

$$\frac{-d[Fe(III)]}{dt} = \frac{k[Fe(III)][I^-]^2}{1 + k'[Fe(II)]/[Fe(III)]}$$

Inner sphere coordination of the halide ion would seem likely in reactions of this type. This is apparently the case for the cerium(IV) oxidations but, although for many years the iron(III) reaction was considered to proceed *via* the iodopentaaquairon(III) ion, recent evidence suggests that this species is not of major importance and that the principal reactant is an outer sphere ion association complex.

Significantly, oxidation with the inert complex hexacyanoferrate-(III) obeys the same rate law. The suggested mechanism for the aqua ion is represented

$$Fe(III) + I^- \rightleftharpoons Fe(III), I^{2+} \qquad 132$$

$$Fe(III), I^{2+} + I^- \rightarrow Fe(II) + I_2^{\cdot -} \qquad 133$$

$$Fe(III) + I_2^{\cdot -} \rightarrow Fe(II) + I_2 \qquad 134$$

The vanadium(V) oxidation of iodide ion

This oxidation provides an example where a two electron step is observed and where complications arise as a result of the presence of an oxygen dependent pathway. As the oxygen dependence is absent in several oxidations where the radical anion $I_2^{\cdot -}$ is implicated, and would not be reasonable for the two electron pathway involving oxoiodic(I) acid as product, this behaviour is due to the vanadium. The oxygen dependent pathway involves formation of a vanadium(V) peroxo complex O_2VO^+. The suggested mechanism (Scheme 4) implies that coordination of iodide ion renders the vanadium(V) more susceptible to electrophilic attack by oxygen.

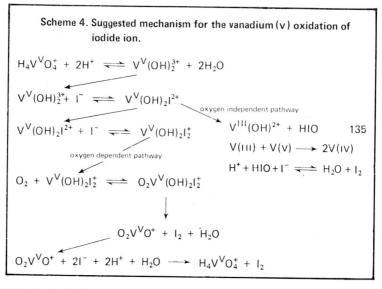

Scheme 4. Suggested mechanism for the vanadium(v) oxidation of iodide ion.

The bismuth(V) oxidation of iodide ion

$$Bi(v) + 2I^- \rightarrow Bi(III) + I_2 \qquad 136$$

By contrast with the examples considered so far the strong oxidant bismuth(v) in dilute acid oxidises iodide ion at a rate independent of iodide ion concentration and first order in bismuth(v). The behaviour is identical to the bismuth(v) oxidation of chloride, bromide, thiocyanate and hexachloroiridate(III). Apparently, the rate determining step involves reaction of bismuth(v) with the solvent (water) or coordinated solvent.

The oxidation of some organic species

The range of organic substrates which have been examined is very wide. Alcohols, aldehydes, ketones, carboxylic acids, phenols, aromatic amines and unsaturated compounds have been particularly well investigated. The principal metal ion oxidants which have been examined are V(v), Cr(vi), Mn(vii), Mn(iii), Fe(iii), Co(iii), Cu(ii), Os(viii), Pd(ii), Ag(i), Hg(ii), Ce(iv), Tl(iii) and Pb(iv).

Not surprisingly, the range of behaviour possible in this class of reactions is considerable. Mechanisms are often of the inner sphere type (Mn(iii), V(v) and Ce(iv)) involving complex formation but are not necessarily so (Fe(CN)$_6^{3-}$). Oxidants involving one electron transfers naturally yield organic free radicals which subsequently open the door to a very wide range of products. The elucidation of a *detailed* mechanism for this type of reaction is thus often beyond present capabilities.

The copper(II) oxidation of reducing sugars in alkaline solution

This reaction in 2,3-dihydroxybutanedioate (tartrate) media (Fehling's solution) or 2-hydroxypropane-1,2,3-tricarboxylate (citrate) media (Benedict's solution) has been widely used for the determination of sugars. The rate of oxidation is independent of the copper(II) concentration. The behaviour in the initial stages of the reaction is described by

$$-\frac{d[Cu(II)]}{dt} = k[S][OH^-] \qquad (S = \text{reducing sugar})$$

The rate determining step is assumed to involve slow formation of the ene-1,2-diol. The suggested mechanism is represented

$$S + OH^- \underset{}{\overset{slow}{\rightleftharpoons}} \begin{array}{c} H \\ | \\ C-O^- \\ \| \\ C-OH \\ | \\ R \end{array} + H_2O \qquad 137$$

$$Cu(II) + \begin{array}{c} H \\ | \\ C-O^- \\ \| \\ C-OH \\ | \\ R \end{array} \underset{}{\overset{fast}{\rightleftharpoons}} \begin{array}{c} H \quad O \\ \diagdown C \diagup \\ \| \quad \diagdown Cu \\ C \diagup \diagdown O \\ R \end{array} + H^+ \longrightarrow Cu_2O + \text{products} \quad 138$$

Deviation from the stated behaviour in the latter stages of the reaction results from involvement of dissolved oxygen.

The chromium(VI) oxidation of alcohols

Chromium(VI) may oxidise primary alcohols to aldehydes and then further to carboxylic acids. Little oxidation to the carboxylic acid is expected if the alcohol is present in excess. Secondary alcohols are oxidised to ketones.

For the oxidation of ethanol (typical of primary or secondary alcohols) the rate equation observed is

$$-\frac{d[Cr(VI)]}{dt} = k_1[H^+]^2[Cr(VI)][C_2H_5OH] + k_2[H^+]^3[Cr(VI)][C_2H_5OH]$$

The chromium(VI) reactant is assumed to be tetraoxochromic(VI) acid (H_2CrO_4) which is in rapid equilibrium with a chromium(VI)–ethanol complex.

$$HCrO_4^- + H^+ \rightleftharpoons H_2CrO_4 \qquad 139$$

$$H_2CrO_4 + CH_3CH_2OH \rightleftharpoons CH_3CH_2O\overset{O}{\underset{\|}{\underset{O}{C}rOH}} + H_2O \qquad 140$$

The rate determining step is considered to be a two electron transfer yielding chromium(IV).

$$CH_3CH_2O\overset{\overset{O}{\|}}{\underset{\underset{O}{\|}}{C}}rOH \rightarrow CH_3\overset{\overset{O}{\|}}{C}H + Cr(IV) \qquad 141$$

Apparently, free radicals are not involved in the reaction, the fate of the chromium(IV) is thus assumed not to be a one electron oxidation of ethanol but the reduction of chromium(VI).

$$Cr(VI) + Cr(IV) \rightarrow 2Cr(V) \qquad 142$$

The chromium(V) produced then oxidises further ethanol

$$Cr(V) + CH_3CH_2OH \rightarrow CH_3\overset{\overset{O}{\|}}{C}H + Cr(III) + 2H^+ \qquad 143$$

presumably through an intermediate analogous to that proposed for the initial chromium(VI) oxidation.

The rate of oxidation is to some extent dependent on the actual acid used. For example, the oxidation is appreciably slower in hydrochloric acid as compared to tetraoxosulphuric(VI) acid. This is attributed to chlorotrioxochromate(VI) (CrO_3Cl^-) replacing the hydrogentetraoxochromate(VI) ($HCrO_4^-$) and sulphatotrioxochromate(VI) ($CrO_3OSO_3^{2-}$) present in the tetraoxosulphuric(VI) acid solution.

The mechanism of oxidation of secondary alcohols appears analogous to that for primary alcohols. Further oxidation of aldehydes to carboxylic acids is also similar, for example in the case of ethanal the mechanism

$$CH_3CHO + H_2CrO_4 \rightleftharpoons CH_3\underset{\underset{H}{|}}{\overset{\overset{OH}{|}}{C}}-OCrO_3H \qquad 144$$

$$CH_3\underset{\underset{H}{|}}{\overset{\overset{OH}{|}}{C}}-OCrO_3H \rightarrow CH_3COOH + Cr(IV) \qquad 145$$

is suggested.

Some manganese(VII) oxidations

Solutions of the tetraoxomanganate(VII) (permanganate) ion will oxidise virtually any organic molecule but the rate varies considerably with the substrate and with the pH of the medium. Unlike chromium(VI) manganese(VII) is a strong oxidant in both acidic and basic solution. The reduction products may be manganese(II) (acidic solution), manganese(IV) oxide (neutral or alkaline solution) or the tetraoxomanganate(VI) ion (strongly alkaline solution). The oxidations may involve direct reaction of the tetraoxomanganate(VII) ion (alkene dihydroxylation) or proceed indirectly (oxidation of ethanedioate) *via* other oxidation states of manganese. In both cases the reactions are complicated by the subsequent fate of the lower oxidation state species. Some illustrative examples are given.

Alkene dihydroxylation
This reaction, useful in the synthesis of diols,* involves direct reaction of the tetraoxomanganate(VII) ion with the formation of a cyclic ester. Good yields of the diol are obtained only in basic solution. The proportion of the appropriate α-hydroxy ketone produced increases with increasing acidity. It may be demonstrated that the α-hydroxy ketone is produced by degradation of the diol, hence a common intermediate is responsible for production of both the diol and the α-hydroxy ketone. The initial stages of the mechanism are represented

$$\begin{array}{c}H\\ C\\ \parallel\\ C\\ H\end{array} + MnO_4^- \longrightarrow \begin{array}{c}H-C-O\\ \diagdown\\ MnO_2^-\\ \diagup\\ H-C-O\end{array} \xrightarrow{H_2O} \begin{array}{c}H-C-OH\\ \\ H-C-OMn^V\end{array}$$

$$\begin{array}{c}H-C-OH\\ \\ C=O\end{array} \xleftarrow{H_2O} \begin{array}{c}H-C-OH\\ \\ H-C-OMn^{VI}\end{array} \xleftarrow{Mn(VII)} \begin{array}{c}H-C-OH\\ \\ H-C-OH\end{array} \xleftarrow{OH^-}$$

146

The competition between hydroxide ion and tetraoxomanganate(VII) ion for the intermediate manganese(V) ester accounts for the decrease in proportion of the diol obtained with increasing acidity.

The oxidation of alcohols
The tetraoxomanganate(VII) oxidation of primary and secondary alcohols in basic solution is generally described by the rate law

$$\frac{-d[MnO_4^-]}{dt} = k[\text{alcohol}][MnO_4^-][OH^-]$$

---

* In practice osmium(VIII) oxide (OsO_4) is more frequently used. The mechanism appears analogous to that for the tetraoxomanganate(VII) ion.

For oxidation of diphenylmethanol ((C_6H_5)$_2$CHOH) the product diphenylmethanone (benzophenone) ((C_6H_5)$_2$CO) contains no oxygen derived from the tetraoxomanganate(VII). Additionally, when (C_6H_5)$_2$CDOH replaces (C_6H_5)$_2$CHOH it is observed that the rate of oxidation is decreased by a factor of ca 6.6 at 25 °C. Apparently, the reaction involves hydride ion transfer from the alkoxide ion to tetraoxomanganate(VII).

$$(C_6H_5)_2CHOH + OH^- \rightleftharpoons C_6H_5CHO^- + H_2O \qquad 147$$
$$(C_6H_5)_2CHO^- + MnO_4^- \rightarrow (C_6H_5)_2CO + HMn^VO_4^{2-} \qquad 148$$
$$\text{rate determining}$$

This mechanism appears general for the tetraoxomanganate(VII) oxidation of alcohols in basic solution

The oxidation of ethanedioate

$$2MnO_4^- + 5C_2O_4^{2-} + 16H^+ \rightarrow 2Mn(II) + 8H_2O + 10CO_2 \qquad 149$$

This reaction is well known for its analytical application in the standardisation of potassium tetraoxomanganate(VII) and in the determination of ethanedioate ion. The reaction is often quoted as an example of autocatalysis. The oxidation is interesting as it is clear that the tetraoxomanganate(VII) ion is not the manganese species oxidising the ethanedioate. It is considered that the initial step involves reaction between managnese(II) and tetraoxomanganate(VII) hence the autocatalysis by manganese(II) and the fact that the reaction is inhibited by fluoride ion (apparently through coordination decreasing the availability of manganese(II)). The initial step is represented

$$\text{Mn(VII)} + \text{Mn(II)} \rightarrow \text{Mn(III)} + \text{Mn(VI)}$$

considered as

$$MnO_4^- + Mn^{II}C_2O_4 \rightarrow Mn^{VI}O_4^{2-} + Mn^{III}C_2O_4^+ \qquad 150$$

then in the absence of appreciable manganese(II) the process is continued

$$\text{Mn(VI)} + C_2O_4^{2-} \rightarrow \text{Mn(IV)} + 2CO_2 \qquad 151$$
$$2\text{Mn(IV)} + C_2O_4^{2-} \rightarrow 2\text{Mn(III)} + 2CO_2 \qquad 152$$

or if manganese(II) is available

$$\text{Mn(VI)} + \text{Mn(II)} \rightarrow 2\text{Mn(IV)} \qquad 153$$
$$\text{Mn(IV)} + \text{Mn(II)} \rightarrow 2\text{Mn(III)} \qquad 154$$

in either case manganese(III) is subsequently destroyed according to the sequence

$$Mn^{III}(C_2O_4)_n^{3-2n} \rightarrow Mn(II) + (n-1)C_2O_4^{2-} + CO_2 + CO_2^{\cdot-} \qquad 155$$
$$Mn^{III}(C_2O_4)_n^{3-2n} + CO_2^{\cdot-} \rightarrow Mn(II) + nC_2O_4^{2-} + CO_2 \qquad 156$$

where $n = 1$, 2 or 3.

The reaction is slightly oxygen sensitive apparently through reaction of the free radical $CO_2^{\cdot-}$

$$CO_2^{\cdot-} + O_2 \rightarrow O_2CO_2^{\cdot-} \qquad 157$$

$$O_2CO_2^{\cdot-} + Mn(II) + 2H^+ \rightarrow Mn(III) + H_2O_2 + CO_2 \qquad 158$$

The hydrogen peroxide is then oxidised by any manganese species in oxidation state III to VII. In view of the extreme complexity of this reaction it is perhaps surprising that an analytically useful stoichiometry is obtained.

Oxidative addition

This term is applied to a particular class of reactions generally involving low spin transition metal complexes (usually of d^7, d^8, or d^{10} configuration), where oxidation occurs with a simultaneous increase in the coordination number of the metal ion concerned. (The reverse of this process is termed reductive elimination.) These reactions have received considerable attention in view of their great synthetic and catalytic utility. The species reduced are frequently, though not necessarily, organic compounds.

The possibilities are illustrated by some reductions involving the pentacyanocobaltate(II) ion and some iridium(I) species. Oxidation of the d^7 species $Co(CN)_5^{3-}$ requires a one electron transfer. The cobalt(III) product (d^6) is invariably six coordinate. Two distinct types of behaviour are observed

$$\frac{-d[Co(CN)_5^{3-}]}{dt} = k[Co(CN)_5^{3-}][S]$$

$$\frac{-d[Co(CN)_5^{3-}]}{dt} = k[Co(CN)_5^{3-}]^2[S]$$

where S is the substrate reduced.

The former is illustrated by the reaction with iodomethane, iodine(I) cyanide, hydroxylamine or hydrogen peroxide. These reactions involve free radicals, for example in the case of iodomethane the mechanism is represented

$$Co(CN)_5^{3-} + CH_3I \rightarrow Co(CN)_5I^{3-} + CH_3^{\cdot} \quad \text{rate determining} \quad 159$$

$$Co(CN)_5^{3-} + CH_3^{\cdot} \rightarrow Co(CN)_5CH_3^{3-} \qquad 160$$

The latter is observed in the reaction with hydrogen and is thought to proceed through a concerted mechanism involving formation of two Co–H bonds.

$$2Co(CN)_5^{3-} + H_2 \rightarrow [(CN)_5Co-H_2-Co(CN)_5]^{6-} \rightarrow 2Co(CN)_5H^{3-} \qquad 161$$

The high heat of dissociation of the hydrogen (435 kJ mol^{-1}) is apparently offset by the cobalt–hydrogen bond energy (243 kJ mol^{-1}).

For d^8 species such as iridium(I) four and five coordinate complexes may be characterised. Oxidative addition to the four coordinate species generally exhibits second order behaviour, the rate being retarded by the products. A typical reaction is

$$(C_6H_5)_3P\underset{Cl}{\overset{CO}{>}}Ir\underset{P(C_6H_5)_3}{} + O_2 \underset{k_{-1}}{\overset{k_1}{\rightleftharpoons}} (C_6H_5)_3P\underset{Cl}{\overset{O_2}{>}}Ir\underset{P(C_6H_5)_3}{\overset{III}{-}}CO \qquad 162$$

The observed rate equation is

$$\frac{-d[Ir(I)]}{dt} = k_1[Ir(I)][O_2] - k_{-1}[Ir(III)]$$

The reaction is apparently simple bimolecular and does not involve the solvent.

By contrast, coordinatively saturated five coordinate iridium(I) complexes undergo dissociation prior to the oxidative addition. A typical reaction is

$$(C_6H_5)_3P\underset{P(C_6H_5)_3}{\overset{OC\quad H}{>}}Ir{-}P(C_6H_5)_3 + R_3SiH \longrightarrow (C_6H_5)_3P\underset{P(C_6H_5)_3}{\overset{OC\quad H}{>}}Ir\underset{SiR_3}{\overset{H}{<}} + P(C_6H_5)_3 \qquad 163$$

The observed rate equation is

$$\frac{-d[Ir(I)]}{dt} = \frac{k_1 k_2[Ir(I)][R_3SiH]}{k_{-1}[P(C_6H_5)_3] + k_2[R_3SiH]}$$

and the suggested mechanism

$$IrHCO(P(C_6H_5)_3)_3 \underset{k_{-1}}{\overset{k_1}{\rightleftharpoons}} IrHCO(P(C_6H_5)_3)_2 + P(C_6H_5)_3 \qquad 164$$

$$IrHCO(P(C_6H_5)_3)_2 + R_3SiH \overset{k_2}{\rightarrow} IrH_2COSiR_3(P(C_6H_5)_3)_2 \qquad 165$$

Experimental

The reduction of some cobalt(III) complexes (p 32)

The importance of a bridging ligand in these reactions may be demonstrated by a qualitative investigation of the vanadium(II) and/or iron(II) reduction of some complexes. Suitable examples* are given in Table 11.

---

* For preparation of the complexes see Appendix 1.

Table 11. Complexes.

Complex[a] 10^{-2} M	Bridging ligand	Observation	
		Addition of excess 2×10^{-2} M V(II) in 5×10^{-1} M H$^+$	Addition of excess 5×10^{-1} M Fe(II) in 5×10^{-1} M H$^+$
trans Co(en)$_2$Cl$_2^+$	Cl	rapid reduction green → virtually colourless	rapid reduction
Co(C$_2$O$_4$)$_3^{3-}$	O	rapid reduction green → virtually colourless	rapid reduction
Co(en)$_3^{3+}$	not available	no reduction	no reduction
Co(NH$_3$)$_6^{3+}$	not available	no reduction	no reduction

[a] Obviously a large number of complexes may be investigated. The complexes Co(NH$_3$)$_5$X^{2+} where X (bridging ligand) = N$_3^-$, Cl$^-$ or Br$^-$ are reduced slightly slower than the examples quoted.

Requirements

trans [Co(en)$_2$Cl$_2$]Cl 10^{-2} M (2.85 g dm^{-3}) ⎫
K$_3$Co(C$_2$O$_4$)$_3$.3.5 H$_2$O 10^{-2} M (5.03 g dm^{-3}) ⎬ These solutions should be freshly prepared.
Co(en)$_3$Cl$_3$ 10^{-2} M (3.45 g dm^{-3})
Co(NH$_3$)$_6$Cl$_3$ 10^{-2} M (2.68 g dm^{-3}) ⎭

(NH$_4$)$_2$SO$_4$.FeSO$_4$.6H$_2$O 5×10^{-1} M in 5×10^{-1} M HCl (196.1 g dm^{-3} 5×10^{-1} M HCl).

V(II) approx. 2×10^{-2} M in approx. 5×10^{-1} M HCl, prepared by zinc reduction of *eg* ammonium polytrioxovanadate(v) (NH$_4$VO$_3$) (2.34 g dm^{-3} MHCl).

Kinetic investigation (20–25 °C). The iron(II) reduction of the trans-dichlorobis(ethane-1,2-diamine)cobalt(III) ion.

Requirements

HCl 5×10^{-1} M
(NH$_4$)$_2$SO$_4$.FeSO$_4$.6H$_2$O 5×10^{-1} M in 5×10^{-1} M HCl (196.1 g dm^{-3} 5×10^{-1} M HCl)
trans [Co(en)$_2$Cl$_2$]Cl 5×10^{-2} M in 5×10^{-1} M HCl (14.25 g dm^{-3} 5×10^{-1} M HCl)
This solution must be freshly prepared.
Colorimeter: filter 545–635 nm (Ilford No. 626) or Spectrophotomer $\lambda = 617$ nm.
Record absorbance *vs* time curves for the mixtures indicated in Table 12.

An 'infinity' value for each run is obtained by warming to 50 °C for 1–2 min.

Table 12. Mixtures for investigation.

5×10^{-2} M trans $Co(en)_2Cl_2^+$ in 5×10^{-1} M HCl (cm³)	5×10^{-1} M HCl (cm³)	5×10^{-1} M Fe(II) in 5×10^{-1} M HCl (cm³)
10	2.5	7.5
10	5.0	5.0
10	7.5	2.5

The conditions chosen in this experiment are such that an excess of iron(II) is present. The observed behaviour is pseudo first order. A plot of log $A_t - A_\infty$ vs time for each reaction yields (-2.303 gradient) the first order rate constants which may be shown to be proportional to [Fe(II)]. The second order rate constant (k) from

$$\frac{-d[Co(en)_2Cl_2^+]}{dt} = k[Co(en)_2Cl_2^+][Fe(II)]$$

is thus obtained.

Literature value: $k_{25°C}$ 3.2×10^{-2} dm³ mol⁻¹ s⁻¹.
Typically $k_{25°C}$ 5.0–13.0×10^{-2} dm³ mol⁻¹ s⁻¹.

Investigation of the stoichiometry of some reactions of trioxosulphate(IV), hydrazine and hydroxylamine

Reactions involving the above reductants exhibit a variable stoichiometry in instances where one electron transfers are involved (p 47). Some reactions suitable for investigation are shown in Table 13. Determination of the stoichiometries is conveniently carried out volumetrically. The variations, indicative of competition for a

Table 13. Mixtures suitable for investigation.

Titrand (25 cm³ portions)	Titrant	Observed stoichiometry M:reductant
10^{-1} M Ce(IV) in 1 M H_2SO_4	10^{-2} M Na_2SO_3[a]	1.43:1
2×10^{-2} M MnO_4^- in 1 M H_2SO_4	10^{-2} M Na_2SO_3[a]	0.30:1
2×10^{-2} M $N_2H_5^+$ in 1 M H_2SO_4	10^{-1} M Ce(IV) in 1 M H_2SO_4[b]	1.44:1
2×10^{-2} M $N_2H_5^+$ in 1 M H_2SO_4	10^{-2} M MnO_4^- in 1 M H_2SO_4[b]	0.30:1
2×10^{-2} M NH_3OH^+ in 1 M H_2SO_4	10^{-1} M Ce(IV) in 1 M H_2SO_4[b]	1.68:1
2×10^{-2} M NH_3OH^+ in 1 M H_2SO_4	2×10^{-2} M MnO^- in 1 M H_2SO_4[b]	0.64:1

[a] If the titration is reversed 10 cm³ 1 M H_2SO_4 must be added to the 25.0 cm³ portion of trioxosulphate(IV).
[b] Titration carried out at 60–70 °C. These oxidations, in common with the chromium(VI) oxidation of hydrazine and the vanadium(V) oxidation of hydroxylamine, proceed at a suitable rate for colorimetric investigation of the kinetics at 20–25 °C.

REDOX REACTIONS INVOLVING METAL IONS

free radical intermediate, may be illustrated by reversing the titration or by varying the concentration of reactants.

The iron(II) induced chromium(VI) oxidation of iodide ion (p 41)

Requirements

K_2CrO_4 2×10^{-3} M in 10^{-1} M H_2SO_4 (0.39 g dm^{-3} 10^{-1} M H_2SO_4)
KI 10^{-2} M (1.66 g dm^{-3})
$(NH_4)_2SO_4.FeSO_4.6H_2O$ 10^{-3} M in 10^{-1} M H_2SO_4 (0.39 g dm^{-3} 10^{-1} M H_2SO_4)
$(NH_4)_2SO_4.FeSO_4.6H_2O$ 5×10^{-2} M in 10^{-1} M H_2SO_4 (19.61 g dm^{-3} 10^{-1} M H_2SO_4)
$NH_4Fe(SO_4)_2.12H_2O$ 10^{-2} M in 10^{-1} M H_2SO_4 (4.82 g dm^{-3} 10^{-1} M H_2SO_4)
H_2SO_4 10^{-1} M
$Na_2S_2O_3.5H_2O$ 10^{-2} M (2.48 g dm^{-3})
1 per cent starch.

Qualitative investigation

(i) The chromium(VI)–iodide ion reaction

To 25 cm^3 of a solution 2×10^{-3} M with respect to chromium(VI) and 10^{-1} M with respect to tetraoxosulphuric(VI) acid add a solution of 25 cm^3 10^{-2} M potassium iodide and 10 cm^3 10^{-1} M tetraoxosulphuric(VI) acid.
The reaction is slow.

(ii) The chromium(VI)–iron(II) reaction

Repeat (i) but replace the 25 cm^3 10^{-2} M potassium iodide with 25 cm^3 distilled water and the 10 cm^3 10^{-1} M tetraoxosulphuric(VI) acid with 10 cm^3 of a solution 10^{-2} M with respect to iron(II) and 10^{-1} M with respect to tetraoxosulphuric(VI) acid.
The reaction is fast.

(iii) The induced oxidation of iodide ion

Repeat (i) but replace the 10 cm^3 10^{-1} M tetraoxosulphuric(VI) acid with 10 cm^3 of a solution 10^{-2} M with respect to iron(II) and 10^{-1} M with respect to tetraoxosulphuric(VI) acid.
The reaction is fast.

(iv) The iron(III)–iodide ion reaction (*see also* p 57).

To 10 cm^3 of a solution 10^{-2} M with respect to iron(III) and 10^{-1} M with respect to tetraoxosulphuric(VI) acid and 25 cm^3 10^{-1} M tetraoxosulphuric(VI) acid add 25 cm^3 10^{-2} M potassium iodide.
The reaction is slow.

Notes

To enable direct comparison of the rates the acidity has been maintained constant throughout these experiments. The chromium-(VI) is considered to be present as hydrogen tetraoxochromate(VI) $(HCrO_4^-)$ or sulphatotrioxochromate(VI) $(CrO_3OSO_3^{2-})$. Starch is a convenient indicator for the reactions involving iodine production. V(IV), U(IV), Cr(II) and $Fe(CN)_6^{4-}$ also behave as inductors.

Determination of the stoichiometry of the iron(II)–chromium(VI)–iodide ion reaction

To establish the stoichiometry of this reaction, the competition of iron(II) for the intermediate responsible for oxidation of the iodide ion must be kept to a minimum. This may be achieved by using a high iodide ion:iron(II) ratio. Attention must, however, be paid to the effect of a high concentration of iodide ion on the chromium(VI)–iodide ion and iron(III)–iodide ion reactions under the same conditions. The stoichiometry is obtained by determining the iodine produced using a fixed concentration of chromium(VI) and varying ratios of iron(II):iodide ion. Conditions and typical results are shown in Table 14.

Table 14. Determining stoichiometry of iron(II)–chromium(VI)–iodide ion.

Reactants[a]				Analysis		
2×10^{-3} M Cr(VI) in 10^{-1} M H_2SO_4 (cm³)	10^{-2} M KI (cm³)	5×10^{-2} M Fe(II) in 10^{-1} M H_2SO_4 (cm³)	10^{-1} M H_2SO_4 (cm³)	Titre[b] 10^{-2} M $S_2O_3^{2-}$ (cm³)	Mole ratio I⁻:Fe(II) used	Mole ratio I⁻:Fe(II) reacted
25	25	20	0	4.5	0.25	0.42
25	25	15	5	5.0	0.33	0.50
25	25	10	10	6.1	0.50	0.68
		10^{-2} M Fe(II) in 10^{-1} M H_2SO_4 (cm³)				
25	25	20	0	7.5	1.25	1.00
25	25	15	5	8.5	1.66	1.31
25	25	10	10	9.0	2.5	1.50
25	25	5	15	8.8	5.0	1.75
25	25	3	17	5.4	8.3	1.8
25	25	2	18	4.2	12.5	2.1
25	25	1	19	1.9	25	1.9
		10^{-3} M Fe(II) in 10^{-1} M H_2SO_4 (cm³)				
25	25	2.5	17.5	0.5	250	2.0

[a] The chromium(VI) should be added last.
[b] The titration must be carried out as soon as the solutions are mixed to minimise the iodine produced from the slow chromium(VI)–iodide ion reaction.

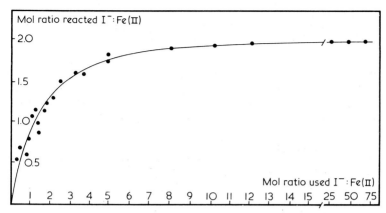

FIG. 4. Mol ratio plot for the iron(II) induced chromium(VI) oxidation of iodide ion (*see* also p 41).

The results are not temperature sensitive over the range 0–25 °C. Above 30 °C the chromium(VI)–iodide ion reaction becomes significant. The data are conveniently evaluated graphically (*Fig. 4*). The competition between iron(II) and iodide ion for the intermediate (chromium(V)) may be inferred from a plot of mole iodide ion reacted *vs* mole iron(II) used.

Notes

The analogous reaction involving vanadium(IV) in place of the iron(II) is also suitable for investigation. In this case there is virtually no competition from vanadium(IV) for the chromium(V).

The iron(III)–iodide ion reaction (p 55)

Requirements

KI 5×10^{-2} M (8.30 g dm^{-3})
NH$_4$Fe(SO$_4$)$_2$.12H$_2$O 5×10^{-2} M (24.11 g dm^{-3})
(NH$_4$)$_2$SO$_4$.FeSO$_4$.6H$_2$O 5×10^{-2} M (19.61 g dm^{-3})
H$_2$SO$_4$ 1 M
1 per cent starch.

Demonstration

This reaction may be examined qualitatively by mixing 10 cm^3 5×10^{-2} M iron(III), 1 cm^3 distilled water and 10 cm^3 5×10^{-2} M potassium iodide. The retarding effect of iron(II) is demonstrated by replacing the 1 cm^3 water by 1 cm^3 5×10^{-2} M iron(II). Starch is a convenient indicator.

Kinetic investigation (15–25 °C)

Colorimeter: filter 370–535 nm (Ilford No. 621 or No. 622) or Spectrophotometer $\lambda = 454$ nm.

Record absorbance *vs* time curves for the mixtures indicated in Table 15. From the gradients at $t = 0$ it may be shown that the initial rate is proportional to $[I^-]^2$ and to $[Fe(III)]$. Examination of the gradients at t_t enables an investigation of the retarding effect of iron(II). Some typical results are shown in *Fig. 5*.

Table 15. Mixtures.

	5×10^{-2} M KI (cm³)	1 M H_2SO_4 (cm³)	Water (cm³)	5×10^{-2} M Fe(III) (cm³)
A	20	10	0	20
B	20	10	10	10
C	10	10	10	20

The tetraoxomanganate(VII) oxidation of ethanedioate ion (p 50)

This kinetic investigation is well documented.[1] A variation involves the examination of some complex ethanedioates.*

Requirements

$KMnO_4$ 1.5×10^{-2} M (2.37 g dm⁻³)
H_2SO_4 2 M
$Na_2C_2O_4$ 1.5×10^{-2} M (2.01 g dm⁻³)
$K_2Cu(C_2O_4)_2 \cdot 2H_2O$ 5×10^{-3} M (1.71 g dm⁻³)

$K_3Fe(C_2O_4)_3 \cdot 3H_2O$ 5×10^{-3} M (2.45 g dm⁻³) ⎫
$K_3Cr(C_2O_4)_3 \cdot 3H_2O$ 5×10^{-3} M (2.43 g dm⁻³) ⎬ These solutions should be freshly prepared.
$K_3Co(C_2O_4)_3 \cdot 3.5H_2O$ 5×10^{-3} M (2.51 g dm⁻³) ⎭

Kinetic investigation (20–25 °C)

Colorimeter: filter 495–575 nm (Ilford No. 624) or Spectrophotometer $\lambda = 526$ nm.

Record absorbance *vs* time curves for the solutions obtained by adding 2.0 cm³ 1.5×10^{-2} M MnO_4^- to each of the following mixtures:

25 cm³ 1.5×10^{-2} M $Na_2C_2O_4$ and 25 cm³ 2 M H_2SO_4.
25 cm³ 5×10^{-3} M $K_3Fe(C_2O_4)_3 \cdot 3H_2O$ and 25 cm³ 2 M H_2SO_4.
25 cm³ 5×10^{-3} M $K_3Cr(C_2O_4)_3 \cdot 3H_2O$ and 25 cm³ 2 M H_2SO_4.
25 cm³ 5×10^{-3} M $K_3Co(C_2O_4)_3 \cdot 3.5H_2O$ and
25 cm³ 2 M H_2SO_4.
25 cm³ 5×10^{-3} M $K_2Cu(C_2O_4)_2 \cdot 2H_2O$ and 25 cm³ 2 M H_2SO_4.

---

* For preparation see Appendix 1.

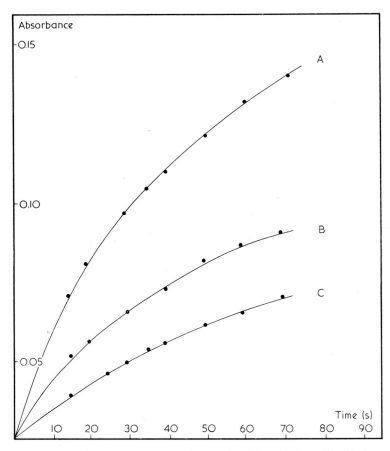

FIG. 5. Absorbance *vs* time curves for the iron(III) oxidation of iodide ion. Conditions as in Table 15. $T = 19\,°C$.

Typical results are shown in *Fig. 6*. The labile complex ions $Fe(C_2O_4)_3^{3-}$ and $Cu(C_2O_4)_2^{2-}$ show no significant difference in behaviour from sodium ethanedioate. The initial rate in the case of the inert chromium(III) and cobalt(III) complexes is, however, limited by dissociation of the complex ions. The autocatalytic effect of manganese(II) may be observed in the case of sodium ethanedioate and the iron and copper complexes.

The chromium(VI) oxidation of ethanol (p 47)[2,3]

This slow reaction is conveniently followed by a sampling procedure using the iodometric method for analysis of chromium(VI).

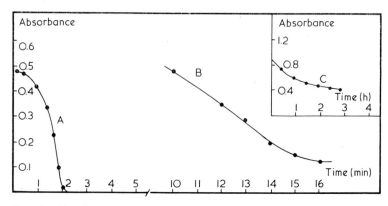

FIG. 6. Absorbance vs time curves for the tetraoxomanganate(VII) oxidation of various ethanedioates. Conditions as in text. $T = 20\,°C$.
(A) $Na_2C_2O_4$, $K_3Fe(C_2O_4)_3 \cdot 3H_2O$ or $K_2Cu(C_2O_4)_2 \cdot 2H_2O$.
(B) $K_3Cr(C_2O_4)_3 \cdot 3H_2O$.
(C) $K_3Co(C_2O_4)_3 \cdot 3.5H_2O$.

Requirements

K_2CrO_4 8×10^{-3} M in 4 M HCl (1.55 g dm^{-3} 4 M HCl)
C_2H_5OH
KI
$Na_2S_2O_3 \cdot 5H_2O$ 10^{-2} M (2.48 g dm^{-3})
1 per cent starch.

Kinetic investigation (20–30 °C)

To 250 cm^3 of a solution 8×10^{-3} M with respect to chromium(IV) and 4 M with respect to hydrochloric acid add 4 cm^3 ethanol. The chromium(VI) concentration is determined by removing 10 cm^3 aliquots (every 5–10 min) adding to each 0.25 g potassium iodide and titrating the liberated iodine with 10^{-2} M sodium trioxothiosulphate(VI) (sodium thiosulphate) using starch as indicator. A plot of log (titre at given time) vs time yields the first order rate constant (-2.303 . gradient).
   Typically $k_{20°C}$ 1–5×10^{-4} s^{-1}.
   The effect of chromium(VI), ethanol and acid may be examined.

References
1. Nuffield Advanced Science, *Chemistry Students Book (II)*, Expt 14.5. Harmondsworth: Penguin, 1970.
2. R. M. Lanes and D. G. Lee, *J. chem. Educ.*, 1968, **45**, 269.
3. M. E. Finlayson and D. G. Lee, *J. chem. Educ.*, 1971, **48**, 473.

5. Some Reactions of the p Block Elements

Reactions of the p block elements may be classified as nucleophilic or electrophilic substitution, free radical, addition, dissociation, isomerisation, exchange and redox reactions. Some redox reactions involving p block elements have been discussed earlier (p 42). There now follows a brief discussion of the various types of reaction together with a more detailed examination of some reactions of the oxides of nitrogen and of clock reactions, principally those involving the halogens.

Addition-dissociation reactions

Electron deficient compounds such as those of the group IIIB elements are capable of forming donor–acceptor complexes, the formation of which may be regarded as an addition and the reverse a dissociation reaction. The boron trifluoride–ammonia adduct provides a typical example

$$BF_3 + :NH_3 \rightarrow F_3B:NH_3 \qquad 166$$

This type of reaction is of some interest because it is accompanied by considerable stereochemical rearrangement, from planar boron trifluoride to the tetrahedral addition product. The mechanism is considered to involve formation of an activated complex which is deactivated by collision with a third body (M)

$$BF_3 + NH_3 \xrightarrow{\text{slow}} F_3BNH_3^* \xrightarrow[\text{fast}]{M} F_3BNH_3 \qquad 167$$

Exchange reactions

Exchange reactions involving p block elements may involve exchange of a given ligand with the solvent or involve ligand exchange between central atoms of the same element or different elements. The exchange of oxygen of oxyanions with the solvent is discussed on p 8. Some reactions involving exchange of halogen atoms are shown in Table 16.

Table 16. Some exchange reactions.

$3SOCl_2 + 2AsF_3 \rightarrow 3SOF_2 + 2AsCl_3$
$BF_3 + AlCl_3 \rightarrow BCl_3 + AlF_3$
$I_2O_5 + 5SF_4 \rightarrow 2IF_5 + 5SOF_2$
$BCl_3 + BF_3 \rightarrow BClF_2 + BFCl_2$
$Si(OCH_3)_4 + SiCl_4 \rightarrow Si(OCH_3)_3Cl + Si(OCH_3)Cl_3$
$Sb(CH_3)_3 + SbCl_3 \rightarrow (CH_3)_2SbCl + CH_3SbCl_2$

A variety of mechanisms are possible for this type of reaction. For example for the exchange

$$MX_3 + MR_3 \rightarrow MX_2R + MXR_2 \qquad 168$$

four suggestions may be made:

1. Electrophilic attack on R (S_E2 reaction)

$$MX_3 \rightleftharpoons MX_2^+ + X^- \qquad 169$$

$$MX_2^+ + MR_3 \rightarrow X_2M \ldots R \ldots MR_2^+ \qquad 170$$
$$\downarrow$$
$$MX_2R + MR^+$$

$$MR_2^+ + X^- \rightarrow MXR_2 \qquad 171$$

2. Nucleophilic attack on M (S_N2 reaction)

$$MX_3 \rightleftharpoons MX_2^+ + X^- \qquad 172$$

$$X^- + MR_3 \rightarrow X \ldots M \ldots R_3^- \qquad 173$$
$$\downarrow$$
$$MXR_2 + R^-$$

$$MX_2^+ + R^- \rightarrow MX_2R \qquad 174$$

3. Free radical attack on M

$$MX_3 \rightarrow X_2M\cdot + X\cdot \qquad 175$$

$$X\cdot + MR_3 \rightarrow XMR_2 + R\cdot \qquad 176$$

$$R\cdot + \cdot MX_2 \rightarrow MX_2R \qquad 177$$

4. Bridge mechanism

$$MX_3 + MR_3 \rightleftharpoons X_2M\underset{R}{\overset{X}{\diagup\!\!\diagdown}}MR_2 \longrightarrow MX_2R + MXR_2 \qquad 178$$

The bridge mechanism is the most frequently encountered, and this has been shown to be the mechanism for the trimethylantimony–antimony(III) chloride exchange. As with all types of reaction, however, additional possibilities are always being discovered. Thus, the rate of the exchange reaction between tetramethoxysilicon and silicon tetrachloride is independent of the concentration of silicon tetrachloride. The reaction

$$Si(OCH_3)_4 + SiCl_4 \rightarrow Si(OCH_3)_3Cl + Si(OCH_3)Cl_3 \qquad 179$$

examined in carbon tetrachloride is acid catalysed. The suggested mechanism is represented

$$Si(OCH_3)_4 + HCl \rightleftharpoons Si(OCH_3)_4 \cdot HCl \qquad 180$$

$$Si(OCH_3)_4 + Si(OCH_3)_4 \cdot HCl \rightleftharpoons Si(OCH_3)_3Cl + Si(OCH_3)_4 + CH_3OH \qquad 181$$

$$CH_3OH + SiCl_4 \rightleftharpoons Si(OCH_3)Cl_3 + HCl \qquad 182$$

The role of trace water in producing hydrogen chloride *via* hydrolysis of silicon tetrachloride is appreciated.

Redox reactions

These reactions involving p block elements usually exhibit two electron changes and generally involve atom or group transfer or replacement. They may be considered as redox reactions or in terms of electrophilic or nucleophilic attack of one species on another. Many reactions may equally well be considered as either. Thus the oxochlorate (I)-trioxosulphate(IV) reaction

$$SO_3^{2-} + ClO^- \rightarrow SO_4^{2-} + Cl^- \quad\quad 183$$

$$S(IV) + Cl(I) \rightarrow S(VI) + Cl(-1) \quad\quad 184$$

could be considered as a two electron redox reaction or as nucleophilic attack by trioxosulphate(IV) on the oxygen of the oxochlorate(I). It is apparent that much of the solution chemistry in this area is that of oxyanions and, consequently, attention in this section has been concentrated on these species.

It is a particular feature of oxyanion chemistry that the reactions are markedly dependent on the hydrogen ion concentration. (This is true not only for p block oxyanions but also for transition metal oxyanions (*eg* MnO_4^- and CrO_4^{2-}) for which much of this discussion is also applicable.) This behaviour results from the ability of protons to labilise the oxygen atom(s) *via* conversion of an oxide ion to a hydroxide ion and subsequently to water. Thus, the oxygen exchange (with solvent water) for these ions is dependent on the hydrogen ion concentration. The dependence is usually first or second order, the former principally for the larger central atoms and/or lower oxidation states as follows:

$\alpha[H^+]^2$ $SO_4^{2-}, NO_3^-, ClO_3^-, BrO^-, CO_3^{2-}$,

$\alpha[H^+]$ IO_3^-, BrO^-, ClO^-

For a given period in the periodic table the rate of oxygen exchange decreases with increasing charge on the central atom, thus we have the series

$$H_2SiO_4^{2-} > HPO_4^{2-} > SO_4^{2-} > ClO_4^-$$

For elements in the same group in the periodic table the rate of oxygen exchange increases with increasing size of the central metal atom. Thus we have the series

$$IO_3^- > BrO_3^- > ClO_3^-$$

Similarly for oxyanions of the same element the rate of oxygen exchange increases with decreasing charge on the central atom and hence we have the series

$$ClO^- > ClO_2^- > ClO_3^- > ClO_4^-$$

Not surprisingly, the rates of oxygen exchange correlate well with the rates for oxidations involving these species. Thus it may be demonstrated that the oxidation of iodide ion by the various oxochlorate anions follows the order

$$ClO^- > ClO_2^- > ClO_3^- > ClO_4^-$$

The oxidation with tetraoxochlorate(VII) is non existent at room temperature as befits a half life for the oxygen exchange of ca 100 y at 25 °C.

Mechanistic details of these exchange processes are generally not available; some suggestions may, however, be made. In the case of tetraoxosulphate(VI) carbonate and trioxochlorate(V) the oxides SO_3, CO_2 and ClO_2^- are stable and a mechanism of the type

$$H_2SO_4 \rightleftharpoons SO_3 + H_2O \quad \text{slow} \qquad 185$$

$$SO_3 + H_2O^* \rightleftharpoons H_2SO_4^* \quad \text{fast} \qquad 18$$

may be proposed.

For trioxonitrate(V) ion the nitryl cation (NO_2^+) is known and the mechanism

$$HNO_3 + H^+ \rightleftharpoons H_2NO_3^+ \qquad 186$$

$$H_2NO_3^+ \rightleftharpoons H_2O + NO_2^+ \quad \text{rate determining} \qquad 187$$

$$NO_2^+ + H_2O^* \rightarrow H_2NO_3^{+*} \qquad 188$$

$$H_2NO_3^{+*} \rightleftharpoons HNO_3^* + H^+ \qquad 189$$

may be suggested.

Some reactions of the oxides of nitrogen

There are six well characterised oxides of nitrogen (Table 17). In addition the free radical $\cdot NO_3$ has been detected spectroscopically and implicated in numerous reactions of the other oxides. The

Table 17. Oxides of nitrogen.

Oxide	Significant equilibria	Comments
N_2O		relatively unreactive
$\cdot NO$	$2 \cdot NO \rightleftharpoons N_2O_2$	oxidised by atmospheric oxygen
N_2O_3	$N_2O_3 \rightleftharpoons \cdot NO_2 + \cdot NO$	appreciably dissociated in the gas phase
$\cdot NO_2$	$2 \cdot NO_2 \rightleftharpoons N_2O_4$	
N_2O_4	$N_2O_4 \rightleftharpoons 2 \cdot NO_2$	appreciably dissociated in the gas phase
N_2O_5	$N_2O_5 \rightleftharpoons \cdot NO_3 + \cdot NO_2$	

majority of reactions of the oxides of nitrogen have been examined in the gas phase. Some, however, including the decomposition of dinitrogen pentoxide have been examined in an inert solvent. Some of the mechanistic possibilities are illustrated by consideration of some reactions of particular oxides.

Dinitrogen pentoxide

The reactions of this oxide all appear to involve $\cdot NO_3$ with which the dinitrogen pentoxide is in equilibrium. The equilibrium may be confirmed by observation of the exchange of ^{15}N labelled nitrogen dioxide

$$N_2O_5 + {}^{15}\cdot NO_2 \rightleftharpoons \cdot NO_2 + \cdot NO_3 + {}^{15}\cdot NO_2 \rightleftharpoons {}^{15}N_2O_5 + \cdot NO_2 \qquad 190$$

Thermal decomposition

$$2N_2O_5 \rightarrow 2N_2O_4 + O_2 \qquad 191$$

This decomposition has been examined in the gas phase and in solvents such as carbon tetrachloride and nitromethane. In all cases the behaviour is described by the rate equation

$$\frac{-d[N_2O_5]}{dt} = k[N_2O_5]$$

The decomposition cannot, however, be considered as an elementary unimolecular reaction as it is apparent that a transition state involving one molecule of dinitrogen pentoxide cannot immediately yield the observed products. The accepted mechanism is represented

$$N_2O_5 \underset{k_{-1}}{\overset{k_1}{\rightleftharpoons}} \cdot NO_3 + \cdot NO_2 \qquad \text{rapidly established} \qquad 192$$

$$\cdot NO_3 + \cdot NO_2 \overset{k_2}{\rightarrow} \cdot NO_2 + O_2 + \cdot NO \qquad \text{rate determining} \qquad 193$$

$$\cdot NO + \cdot NO_3 \overset{k_3}{\rightarrow} 2\cdot NO_2 \qquad 194$$

$$2\cdot NO_2 \rightleftharpoons N_2O_4 \qquad 195$$

which on application of the steady state $(d[\cdot NO_3]/dt = 0, d[\cdot NO]/dt = 0)$ theory yields the rate equation

$$\frac{-d[N_2O_5]}{dt} = \frac{2k_1k_2[N_2O_5]}{k_{-1} + 2k_2}$$

In agreement with the above mechanism nitrogen oxide is observed to increase the rate of decomposition. This effect is apparently a result of the removal of $\cdot NO_3$. The stoichiometry is then represented

$$N_2O_5 + \cdot NO \rightarrow 3\cdot NO_2 \qquad 196$$

and the suggested mechanism is

$$N_2O_5 \rightleftharpoons \cdot NO_3 + \cdot NO_2 \qquad 192$$

$$\cdot NO + \cdot NO_3 \rightarrow 2 \cdot NO_2 \qquad 194$$

$$2 \cdot NO_2 \rightleftharpoons N_2O_4 \qquad 195$$

Nitrogen dioxide
Thermal decomposition

$$2 \cdot NO_2 \rightarrow 2 \cdot NO + O_2 \qquad 197$$

This reaction has been investigated in the gas phase at 227–377 °C. The behaviour is roughly described by the rate equation

$$\frac{-d[\cdot NO_2]}{dt} = k[\cdot NO_2]^2$$

The reaction is initially rapid but beyond *ca* 10 per cent decomposition obeys the stated equation well. The rapid initial phase may be eliminated by the addition of nitrogen oxide. This behaviour is explained by considering two concurrent pathways. The suggested mechanism is represented

$$2 \cdot NO_2 \rightarrow 2 \cdot NO + O_2 \qquad 198$$

$$2 \cdot NO_2 \rightarrow \cdot NO_3 + NO \qquad 199$$

$$\cdot NO + \cdot NO_3 \rightarrow 2 \cdot NO_2 \qquad 194$$

$$\cdot NO_3 + \cdot NO_2 \rightarrow \cdot NO_2 + O_2 + \cdot NO \qquad 193$$

Reaction with fluorine

$$2 \cdot NO_2 + F_2 \rightarrow 2NO_2F \qquad 200$$

This reaction in the gas phase at 27–97 °C is described by the rate equation

$$\frac{-d[\cdot NO_2]}{dt} = k[\cdot NO_2][F_2]$$

The suggested mechanism is represented

$$\cdot NO_2 + F_2 \rightarrow NO_2F + \cdot F \quad \text{rate determining} \qquad 201$$

$$\cdot NO_2 + \cdot F \rightarrow NO_2F \qquad 202$$

Nitrogen oxide
Reaction with oxygen*

$$2 \cdot NO + O_2 \rightarrow 2 \cdot NO_2 \qquad 203$$

---

* It should be noted that the oxygen molecule is itself a diradical (*ie* contains two unpaired electrons). See, for example, F. A. Cotton and G. Wilkinson, *Advanced inorganic chemistry*, 3rd Edn, p 106. New York: Interscience, 1972.

This reaction has been examined in the gas phase and obeys the rate equation

$$\frac{-d[\cdot NO]}{dt} = k[\cdot NO]^2[O_2]$$

The reaction as is well known (colourless→brown) is rapid at room temperature. The mechanism

$$2\cdot NO \rightleftharpoons N_2O_2 \qquad 204$$

$$N_2O_2 + O_2 \rightarrow 2\cdot NO_2 \qquad 205$$

is suggested.

Evidence for the mechanism has been obtained from isolation of N_2O_2 from nitrogen oxide at $-258\,°C$ and from a study of the effect of temperature on the reaction. The rate is observed to decrease with increasing temperature. This suggests (Le Chatelier's principle) that an exothermic equilibrium is involved such that only the species on the right hand side can react to give the required products. An exothermic equilibrium involving production of the reactive species N_2O_2 would be in agreement with this requirement.

Catalysis involving nitrogen oxide–nitrogen dioxide

Nitrogen dioxide is a good catalyst for a number of gas phase oxidations. The process involves reduction to nitrogen oxide and subsequent oxidation by molecular oxygen. For the oxidation of carbon monoxide (206)

$$2CO + O_2 \rightarrow 2CO_2 \qquad 206$$

the catalysis is represented

$$2(CO + \cdot NO_2 \rightarrow CO_2 + \cdot NO) \qquad 207$$

$$2\cdot NO + O_2 \rightarrow 2\cdot NO_2 \qquad 203$$

Clock reactions

This name is given to a type of reaction where at some time after initial mixing of the reagents the sudden appearance of a product is observed. The term clock arises from the fact that at fixed concentrations and constant temperature the appearance of the product(s) occurs at a definite time after the initial mixing. Reactions of this kind are also referred to as Landolt reactions after Landolt who first observed this effect in a study of the hydrogen trioxosulphate(IV)–trioxoiodate(V) reaction. The reactions are generally of the oxidation–reduction type involving halogens in various oxidation states but acid–base and complexation reactions can also be contrived to show the effect. In this respect it is worth emphasising that a reaction may exhibit this type of behaviour *per se* or be a system deliberately designed to do so (p 71). In either case the behaviour results from feedback of the products or from competitive inhibition of some state en route to the product(s).

Perhaps the best known example where the situation is real as opposed to contrived is the hydrogen trioxosulphate(IV)–trioxoiodate(V) reaction

$$2IO_3^- + 5HSO_3^- + 2H^+ \rightarrow I_2 + 5HSO_4^- + H_2O \qquad 208$$

This reaction in the presence of excess trioxoiodate(V) is characterised by the sudden appearance of iodine at sometime (t) after the initial mixing. The system is represented

$$IO_3^- \xrightarrow{HSO_3^-} I^- \underset{HSO_3^-}{\xrightarrow{IO_3^-}} I_2 \qquad 209$$

and involves three gross reactions

$$IO_3^- + 3HSO_3^- \xrightarrow{\text{slow}} 3HSO_4^- + I^- \qquad 210$$

$$IO_3^- + 5I^- + 6H^+ \rightarrow 3I_2 + 3H_2O \qquad 211$$

$$I_2 + HSO_3^- + H_2O \rightarrow 2I^- + HSO_4^- + 2H^+ \qquad 212$$

The reactions are such that 212 is very much faster than 210 and 211. Thus the iodine produced *via* 211 is removed as fast as it is produced. This behaviour continues until the hydrogen trioxosulphate(IV) is virtually exhausted, reaction 212 then begins to slow down and the iodine concentration (produced *via* 211) rises. Under suitable conditions the iodine may appear very rapidly. The important point is not that the trioxoiodate(V)–iodide reaction 211 is slow but merely that it is slower than the hydrogen trioxosulphate(IV)–iodine reaction 212.

The detailed mechanism of the three gross reactions is not germane to discussion of the clock behaviour but will be considered briefly. In the absence of added acid the induction period (t) is represented

$$t = \frac{k}{[KIO_3]_0[NaHSO_3]_0}$$

indicating that the average rate during this period is proportional to $[IO_3^-]_0$ and $[HSO_3^-]_0$.

The rate law reported for 210 is

$$\frac{-d[HSO_3^-]}{dt} = [H^+][IO_3^-][HSO_3^-]\{k_1 + k_2[SO_3^{2-}]\}$$

The principal pathway is considered to involve three two electron transfers

$$H^+ + HSO_3^- + IO_3^- \rightarrow HSO_4^- + HIO_2 \quad \text{rate determining} \qquad 213$$

$$HIO_2 + HSO_3^- \rightarrow HSO_4^- + HIO \qquad 214$$

$$HIO + HSO_3^- \rightarrow HSO_4^- + I^- + H^+ \qquad 215$$

The trioxoiodate(v)–iodide reaction 211 is very complex and various rate laws involving terms in $[I^-]$, $[I^-]^2$ and $[I^-]^3$ have been proposed. In reasonably acidic solution (pH 3) it is apparent that the reaction is analogous to the trioxobromate(v) reaction[1, 2] with the rate expression

$$\frac{-d[IO_3^-]}{dt} = \frac{1}{3}\frac{d[I_2]}{dt} = k[IO_3^-][I^-][H^+]^2$$

leading to the mechanism

$IO_3^- + 2H^+ + I^- \rightarrow HIO_2 + HIO$ rate determining 216

$HIO_2 + H^+ + I^- \rightarrow 2HIO$ 217

$HIO + H^+ + I^- \rightleftharpoons I_2 + H_2O$ 218

The iodine–hydrogentrioxosulphate(IV) reaction 212 obeys the rate law

$$\frac{-d[I_2]}{dt} = k[HSO_3^-][I_2]$$

which is interpreted as resulting from the mechanism

$HSO_3^- + I_2 \rightleftharpoons HSO_3I + I^-$ 219

$HSO_3I + H_2O \rightleftharpoons HSO_4^- + 2H^+ + I^-$ 220

In contrast to the hydrogen trioxosulphate(IV)–trioxoiodate(V) reaction where the eventual product is continuously removed *via* a feedback mechanism the dioxochlorate(III)–iodide reaction illustrates a situation where the clock behaviour involves two distinct stages and competitive inhibition is responsible for the initial non-existence of the second phase.

The dioxochlorate(III) oxidation of iodine in the presence of excess dioxochlorate(III) is represented

$6ClO_2^- + 4I^- \rightarrow 6Cl^- + 4IO_3^-$ 221

The two phases considered are

$ClO_2^- + 4I^- + 4H^+ \rightarrow 2I_2 + 2H_2O + Cl^-$ 222

and

$5ClO_2^- + 2I_2 + 2H_2O \rightarrow 4IO_3^- + 4H^+ + 5Cl^-$ 223

The oxidation of iodide to iodine is catalysed by iodine and is thus autocatalytic while the oxidation to trioxoiodate(V) is inhibited by iodide and is thus not observed until virtually all the iodide is consumed. Iodine may be regarded as an observable intermediate in the oxidation to trioxoiodate(V). The behaviour is shown in *Fig. 7*.

The suggested mechanism of iodine production is represented

$$H^+ + ClO_2^- \rightleftharpoons HClO_2$$

$$H^+ + I^- + HClO_2 \longrightarrow HClO + HIO \qquad \text{initial rate determining}$$

$$HClO + I^- \longrightarrow HIO + Cl^-$$

$$HIO + I^- + H^+ \rightleftharpoons I_2 + H_2O$$

$$HIO + HClO_2 \longrightarrow HIO_2 + HClO$$

$$2HIO \longleftarrow H^+ + I^- + HIO_2$$

$$HIO + Cl^- \longleftarrow I^- + HClO$$

(224)

The autocatalysis apparently results from the hydrolysis of iodine which causes an increase in the oxoiodic(I) acid concentration. Whatever the subsequent fate of the oxoiodic(I) acid this only serves to produce even more oxoiodic(I) acid via consumption of iodide–dioxochlorate(III). The oxidation of iodine to trioxoiodate(V) is then represented

$$I_2 + H_2O \rightleftharpoons HIO + H^+ + I^- \qquad (225)$$

$$HIO + HClO_2 \rightarrow HIO_2 + HClO \qquad (226)$$

$$HIO + HClO \rightarrow HIO_2 + H^+ + Cl^- \qquad (227)$$

FIG. 7. Absorbance vs time curve for the dioxochlorate(III) oxidation of iodide in the presence of excess dioxochlorate(III). For conditions see p 81.

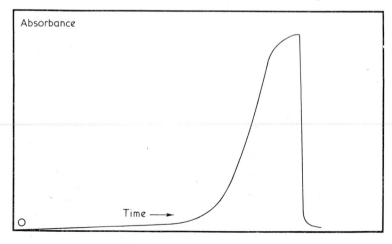

SOME REACTIONS OF THE P BLOCK ELEMENTS

$$HClO_2 + HIO_2 \rightarrow IO_3^- + HClO + H^+ \qquad 228$$

$$HClO + HIO_2 \rightarrow IO_3^- + 2H^+ + Cl^- \qquad 229$$

The two distinct phases result from the competition between iodide and oxochloric(I) acid or dioxochloric(III) acid for oxoiodic(I) acid and dioxoiodic(III) acid.

$$\begin{array}{c} \text{HIO} \begin{array}{c} \xrightarrow{I^-} I_2 \\ \xrightarrow{HClO,\ HClO_2} IO_3^- \end{array} \qquad \text{HIO}_2 \begin{array}{c} \xrightarrow{I^-} \text{HIO} \\ \xrightarrow{HClO,\ HClO_2} IO_3^- \end{array} \end{array} \qquad 230$$

In the presence of appreciable iodide ion, oxidation of the iodine species is not observed, dioxoiodic(III) acid is removed as fast as it is produced. As the iodide ion concentration falls the oxidation is no longer inhibited and the iodine is thus rapidly oxidised.

The methanal–trioxosulphate(IV)–hydrogentrioxosulphate(IV) reaction

This clock reaction illustrates a system where halogens are not involved. The reaction involves a rapid rise in pH on exhaustion of the hydrogentrioxosulphate(IV) (*Fig. 8*). Any pH indicator changing colour in the pH range 7–11 will render the clock effect visible.

The reaction is represented

(*i*) Formation of a hydrogentrioxosulphate(IV) addition compound

$$\underset{H}{\overset{H}{>}}C=O + HSO_3^- \xrightarrow{\text{slow}} \underset{H}{\overset{H}{>}}C\underset{O}{\overset{OH}{<}}S\underset{O^-}{\overset{O}{<}} \qquad 231$$

$$\underset{H}{\overset{H}{>}}C=O + H_2O + SO_3^{2-} \xrightarrow{\text{slow}} \underset{H}{\overset{H}{>}}C\underset{O}{\overset{OH}{<}}S\underset{O^-}{\overset{O}{<}} + OH^- \qquad 232$$

(*ii*) Rapid removal of hydroxide ion formed in 233

$$HSO_3^- + OH^- \rightarrow H_2O + SO_3^{2-} \qquad 233$$

Exhaustion of the hydrogentrioxosulphate(IV) (pH ~ 7) leads to a rapid rise in hydroxide ion concentration as a result of 232.

Design of a clock reaction

With knowledge of a system of the hydrogentrioxosulphate(IV)–trioxoiodate(V) type it is possible to design a large number of systems involving feedback of a reactant through consumption of a

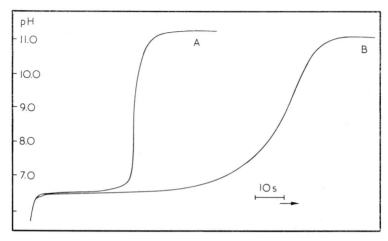

FIG. 8. pH vs time curves for the methanal–trioxosulphate(IV)–hydrogen trioxosulphate(IV) reaction. $T = 22\,°C$. (A) Stirred solution. (B) Unstirred. Conditions as in Table 20.

product by an added reagent. For example the reaction

$$A + B \rightarrow C$$

will exhibit this behaviour if substance D is added to return C to B as fast as it is produced

$$A + B \rightarrow \underset{D}{C}$$

In the presence of excess A the product C is then observed on exhaustion of D.

Interest in these reactions has resulted from the fact that they may be used analytically either for the determination of D or of a catalyst for the reaction. Some examples are given in Table 18.

Table 18. Some analytically useful clock reactions.

A	B	C	D	Analytical application
BrO_3^- [b]	Br^-	Br_2	various phenols	determination of phenols
$S_2O_8^{2-}$	I^-	I_2	$S_2O_3^{2-}$	determination of Cu(II)[a, c]
H_2O_2	I^-	I_2	$S_2O_3^{2-}$	determination of Mo(VI)[a]
H_2O_2 [b]	Br^-	Br_2	ascorbic acid	determination of Cu(II), Fe(III), V(V) and W(VI)[a]
ClO_3^- [b]	Cl^-	Cl_2	ascorbic acid	determination of Mo(VI) and V(V)[a]

[a] As a catalyst for the reaction.
[b] Azo carmine GX is a useful indicator for these reactions.
[c] See also p 87.

Experimental

The trioxoiodate(V)–hydrogentrioxosulphate(IV) reaction (p 68)

Requirements

KIO$_3$ 2×10^{-2} M (4.28 g dm^{-3})
NaHSO$_3$ 2×10^{-2} M (1.90 g Na$_2$S$_2$O$_5$ dm^{-3})
1 per cent starch.

Kinetic investigation (20–30 °C)

Record the time (t) for iodine to be produced in each of the mixtures in Table 19. Add a little starch as indicator. Note the time for the blue starch–iodine colour to appear. The reaction mixture should be stirred if possible.

Table 19 Mixtures.

2×10^{-2} M NaHSO$_3$ (cm^3)	2×10^{-2} M KIO$_3$ (cm^3)	H$_2$O (cm^3)
25	25	50
25	50	25
25	45	30
25	40	35
25	60	15
25	65	10
35	25	40
50	50	0
75	25	0

A plot of time period (t) vs $1/[\text{HSO}_3^-]_0[\text{IO}_3^-]_0$ (*Fig. 9*) (where the subscript $_0$ refers to the initial concentrations) enables confirmation of the relationship

$$t = \frac{k}{[\text{HSO}_3^-]_0[\text{IO}_3^-]_0}$$

Literature value $k_{23°C}$ 3.7×10^{-3} dm^3 mol^{-1} s^{-1}.
Typically $k_{22°C}$ $3-5 \times 10^{-3}$ dm^3 mol^{-1} s^{-1}.

The dioxochlorate(III)–iodide reaction (p 69)

Requirements

KI 5×10^{-4} M (0.083 g dm^{-3})
NaClO$_2$ *ca* 5×10^{-3} M*
pH 6 buffer (2.4 g Na$_2$HPO$_4$ and 8.1 g KH$_2$PO$_4$ dm^{-3})
1 per cent starch.

---

* NaClO$_2$ is generally only available in a purity of *ca* 80 per cent. Allowance should be made for the stated purity. If the solution is to be stored for more than one hour it should be prepared in 10^{-3} M NaOH.

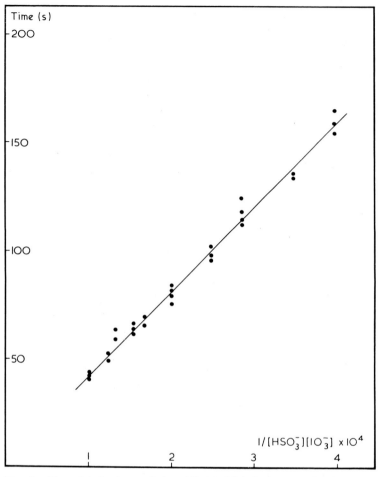

FIG. 9. Plot of induction period *vs* $1/[HSO_3^-]_0[IO_3^-]_0$ for the trioxoiodate(v) oxidation of hydrogen trioxosulphate(IV) in the presence of excess trioxoiodate(v). $T = 22\,°C$. Conditions as in Table 19.

Demonstration

To 10 cm³ 5×10^{-4} M potassium iodide solution add 25 cm³ pH 6 buffer, a few drops of 1 per cent starch and 25 cm³ *ca* 5×10^{-3} M sodium dioxochlorate(III). Note the time for the blue starch–iodine colour to disappear.* The reaction mixture should be stirred if possible.

---

* There does not appear to be a simple relationship between the initial concentrations and the period of iodine production.

The methanal–trioxosulphate(IV)-hydrogentrioxosulphate(IV) reaction (p 71)[3, 4]

Requirements
(A) HSO_3^-/SO_3^{2-} (20 g $Na_2S_2O_5$, 3 g $Na_2SO_3 \cdot 7H_2O$ dm^{-3})
(B) HCHO (65 cm^3 formalin (40 per cent HCHO) dm^{-3})
phenolphthalein, bromophenol blue or thymolphthalein indicator.

Kinetic investigation (20–30 °C)
Record the time (t) for the indicator to change colour in each of the mixtures in Table 20. The reaction mixture should be stirred if possible.

Table 20. Mixtures.[a]

Solution A (cm^3)	Solution B (cm^3)	H$_2$O (cm^3)
25	40	235
25	25	250
25	20	255
25	15	260
40	25	235
60	25	215
75	25	200

[a] A few drops of indicator solution should be added to each mixture.

Under these conditions the rate during the induction period is represented rate = $k[A]^a[B]^b$. A plot of log [rate] (ie 1/time) vs log [A] at fixed [B] yields a graph of gradient a. Similarly a plot of log [rate] vs log [B] at fixed [A] yields b. Solution volumes may be used as a measure of the concentrations of A and B.

$$a = -1 \quad b = +1$$

The reaction may also be followed using a recording pH meter (*Fig. 8*).

References
1. P. G. Ashmore, *Principles of reaction kinetics*, 2nd edn, p 48. London: The Chemical Society, 1973.
2. J. R. Clarke, *J. chem. Educ.*, 1970, **47**, 775.
3. R. L. Barrett, *J. chem. Educ.*, 1955, **32**, 78.
4. T. Cassen, *J. chem. Educ.*, 1976, **53**, 197.

6. Metal Ion Assisted Processes

Metal ions are involved in the promotion of a large number of reactions* varying from the microscopic scale of vital processes within a single plant or animal cell to large scale industrial operations. Reactions which are considered in this chapter are:

1. Reactions which are forced to occur at a limited number of sites through coordination rendering certain sites unavailable for reaction.
2. Reactions in which ligands are stabilised in a fixed tautomeric configuration as a result of coordination.
3. Reactions which are favoured as a result of coordination subjecting a ligand to internal strain.
4. Reactions involving electron redistribution within a ligand as a result of coordination to a positively charged metal ion.
5. Reactions in which the role of the metal ion is organisational in promoting the union of two or more ligands to yield a chelate. Such reactions are commonly referred to as template reactions and generally would not occur in the absence of a metal ion.
6. Insertion reactions.
7. Oligomerisation and polymerisation reactions.
8. Oxidation–reduction reactions in which a metal ion, through its ability to exist in two or more oxidation states, provides a more favourable pathway for a redox reaction.

A large number of actual reactions involve a combination of two or more of the above.

Reactions at a limited number of sites

Donation of a lone pair of electrons to a metal ion prevents involvement of that pair of electrons in further reaction. For example, the condensation of an amine and aldehyde

$$RCHO + NH_2CH_2R' \rightarrow RCH=NCH_2R' + H_2O \qquad 234$$

is inhibited by coordination of the amine group. This behaviour may be exploited by using a metal ion to protect the amine group. The mechanism is admirably illustrated by the copper(II) assisted

---

* In many of these reactions the role of the metal ion is that of a catalyst. A catalyst is considered as a substance which appears as a concentration term (to a positive power) in the rate expression but which does *not* appear in the stoichiometric equation. The catalysed reaction may be regarded as a chain reaction with the catalyst consumed in one step and regenerated in a later one.

synthesis of threonine from aminoethanoic acid and ethanal

$$\begin{array}{c} \text{O}=\text{C}-\text{O} \\ | \quad \quad \text{Cu} \\ \text{CH}_2-\text{N} \quad \text{O}-\text{C}=\text{O} \\ \text{H}_2 \end{array} \begin{array}{c} \text{NH}_2-\text{CH}_2 \\ | \end{array} \xrightleftharpoons{\text{OH}^-} \begin{array}{c} \text{O} \\ \text{C}-\text{O} \\ | \quad \quad \text{Cu} \\ \text{CH}-\text{N} \quad \text{O}-\text{C}=\text{O} \\ \text{H}_2 \end{array} \begin{array}{c} \text{NH}_2-\text{CH}_2 \\ | \end{array} + \text{H}_2\text{O}$$

$\Updownarrow \text{CH}_3\text{CHO}/\text{H}_2\text{O}$ 235

$$\begin{array}{c} \text{O}=\text{C}-\text{O} \\ | \quad \quad \quad \text{Cu} \\ \text{CH}_3-\text{CH}-\text{CH}-\text{N} \quad \text{O}-\text{C}=\text{O} \\ | \quad \quad \quad \text{H}_2 \\ \text{OH} \end{array} \begin{array}{c} \text{NH}_2-\text{CH}_2 \\ | \end{array} + \text{OH}^-$$

In addition to protecting the amine group it appears that the metal ion also facilitates the formation of the enolate ion favouring attack of the aldehyde group. The threonine may be obtained by removal of the copper(II) with, for example, hydrogen sulphide.

Reactions involving stabilisation in a particular configuration

An example of this type of behaviour is afforded by ketones where the keto–enol equilibrium is disturbed on coordination. Pentane-2,4-dione, for instance, exists as *ca* 72 per cent enol form and 28 per cent keto form at room temperature. On coordination effectively 100 per cent is present as the enol form. Other things being equal the metal ion complex thus reacts more rapidly than the free ligand in those instances where the keto–enol equilibrium is rate determining (*eg* halogenation). It should, however, be remembered that the rate of formation of the complex is at least as slow as the rate of enolisation.

Reactions favoured as a result of internal strain

Confirmation of a mechanism as involving strain is difficult. The phenomenon has, however, been suggested in numerous instances and is referred to as a rack mechanism. The increased reactivity apparently results from a lowering of the energy difference between the ground state and the transition state for the reaction.

A reaction where ligand strain is thought to be significant is the metal ion catalysed decomposition of hydrogen peroxide. This reaction involving breakage of the peroxo bond appears to be favoured as a result of strain of the peroxo bond on coordination. A suggested mechanism for the dihydroxotriethenetetramine-iron(III) ion catalysed decomposition of hydrogen peroxide is shown in Scheme 5.

Scheme 5. Catalytic decomposition of hydrogen peroxide by Fe(trien)(OH)$_2^+$.

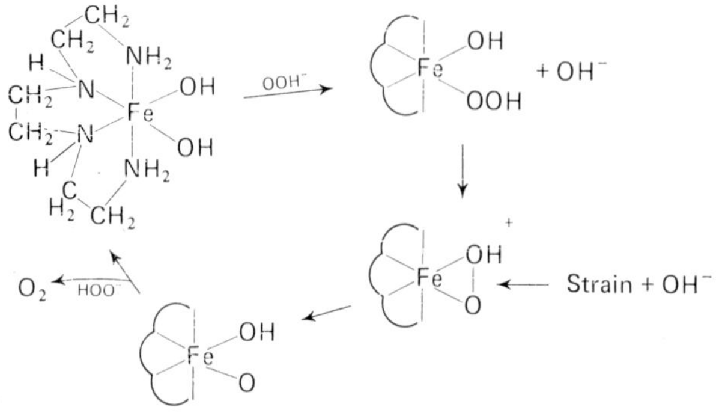

(236)

Reactions involving electron redistribution

Metal ion assisted hydrolyses of various organic species provide examples of this type of mechanism. The reactions are of considerable biochemical importance. Systems involving both labile and inert metal ions have been investigated. It is apparent that the hydrolysis results from electron redistribution within the ligand favouring nucleophilic attack by hydroxide ion or water. The behaviour is illustrated by the M^{2+} assisted hydrolysis of amino acid esters.

The rate of hydrolysis of the ethyl ester of aminoethanoic acid (glycine) is increased by presence of the labile metal ions cobalt(II) and copper(II). The first order rate constant for the hydrolysis approaches a maximum as the metal ion:ester ratio approaches unity which suggests a 1:1 complex as intermediate.

The proposed mechanism is represented

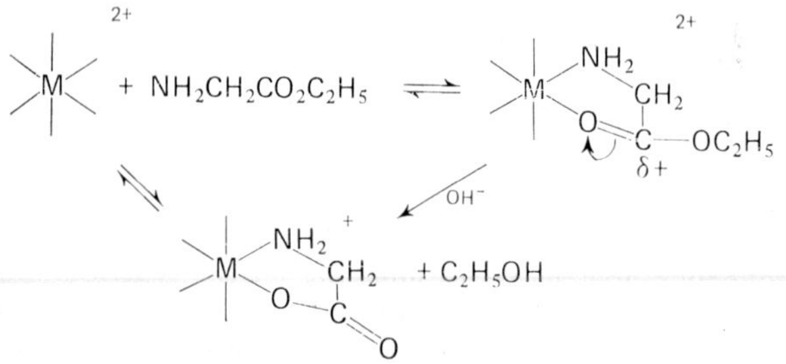

(237)

The implied coordination of the carbonyl group would lead to polarisation thus rendering the carbon atom more susceptible to nucleophilic attack by hydroxide ion. Support for coordination of the carbonyl group is obtained from the infrared spectra of 1:1 metal ion–ester complexes which may be isolated and from the

Scheme 6. Suggested mechanism for the hydrolysis of ethyl aminoethanoate in the complex $Co(en)_2(NH_2CH_2COOC_2H_5)Cl^{2+}$.

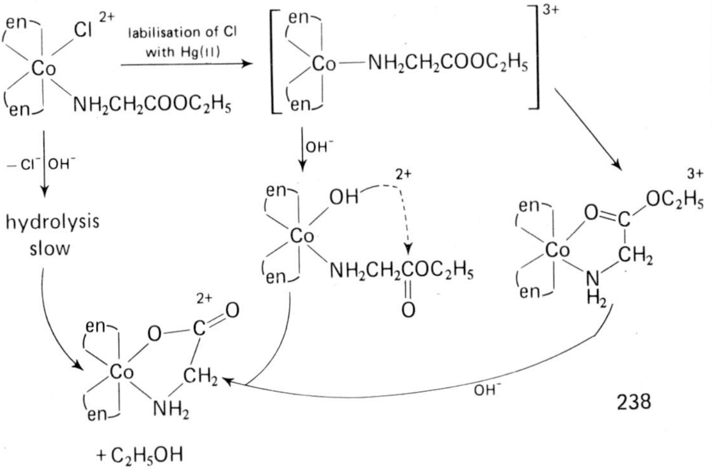

238

analogous inert cobalt(III) complexes. For example, the ester in the complex ion $Co(en)_2NH_2CH_2CO_2C_2H_5Cl^{2+}$ (containing the ester as a unidentate (N bonded) ligand) is hydrolysed only slowly. Labilisation of the coordinated chloride ion using mercury(II), however, renders the hydrolysis rapid, in agreement with with the requirement of bidentate coordination of the ester. Studies with oxygen-18, however, indicate that this is only one of two possible pathways, the second involving attack of a coordinated hydroxide ion (Scheme 6).

Template reactions

In this type of reaction a metal ion fixes ligands favourably for a slow condensation reaction which results in ring closure. The reactions are important in the synthesis of several biochemicals (*eg* porphyrins and corrin) the *in vivo* synthesis of which is incompletely understood. A typical laboratory synthesis is illustrated by the reaction of the tris(ethane-1,2-diamine)nickel(II) ion with dry propanone

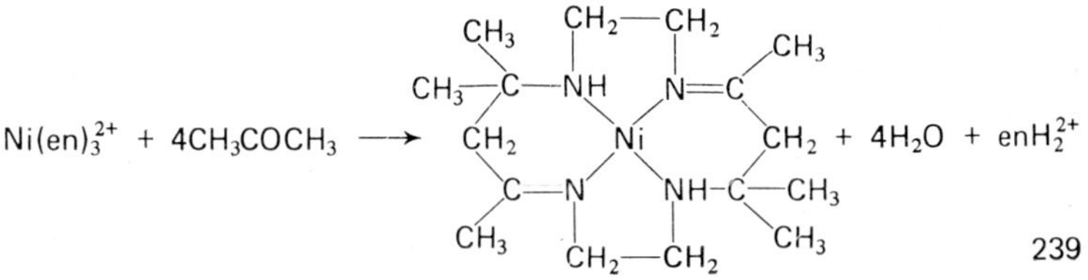

239

Details of the mechanism of this type of reaction have not proved easy to obtain. It is, however, apparent that the reactions involve a series of condensation reactions. The initial stage in the above system is thus thought to be represented as

$$Ni(en)_3^{2+} + 4CH_3COCH_3 \longrightarrow \begin{array}{c} \text{complex structure} \end{array} + 4H_2O + en \qquad 240$$

With the water removed using *eg* anhydrous calcium sulphate, ethane-1,2-diamine then serves as a base enabling further condensation.

$$\begin{array}{c}\text{Ni complex} + NH_2CH_2CH_2NH_2 \longrightarrow \text{Ni complex} + NH_2CH_2CH_2NH_3^+ \end{array}$$

$$\begin{array}{c}\downarrow \\ \text{Ni complex} + NH_2CH_2CH_2NH_2 \\ etc \end{array} \qquad 241$$

Insertion reactions

These combination reactions involve the insertion of a group between a metal ion and ligand. The name insertion may be mechanistically misleading. In the light of mechanistic information the reactions have sometimes been termed migration reactions.

The process is represented

$$ML + L' \rightarrow ML'L \qquad 242$$

where L is the ligand and L' the group interested.

The most common examples involve combination of alkenes, carbon monoxide or sulphur dioxide with alkyl groups. A particular example is

$$CH_3Mn(CO)_5 + CO \rightleftharpoons CH_3COMn(CO)_5 \qquad 243$$

Two mechanisms may be considered for this reaction, the first involving carbon monoxide insertion into the metal–alkyl bond, the second involving migration of the alkyl group to a coordinated carbon monoxide molecule. Using ^{14}CO it has been shown that the carbon monoxide molecule which is 'inserted' (*ie* becomes the

METAL ION ASSISTED PROCESSES

acyl carbonyl) is not derived from external carbon monoxide but is one already coordinated to the metal ion.

Interestingly, the conversion can be effected by ligands other than carbon monoxide, eg

$$CH_3Mn(CO)_5 + P(C_6H_5)_3 \rightarrow CH_3COMn(CO)_4P(C_6H_5)_3 \quad\quad 244$$

Furthermore it may be shown that the incoming group is added in a position *cis* with respect to the acyl group. The mechanism suggested involves migration of the alkyl group and probably proceeds through a three centre transition state

$$\begin{array}{c}CH_3\\|\\Mn\!-\!CO\end{array} \longrightarrow \begin{array}{c}CH_3\\\diagup\;\diagdown\\Mn\!\cdot\!\cdot\!CO\end{array} \longrightarrow M\!-\!C\!\!\begin{array}{c}\diagup CH_3\\\diagdown O\end{array} \quad\quad 245$$

It is, however, evident that there are other possibilities for this type of reaction since carbon monoxide insertion in the analogous reaction of *cis* $CH_3Mn(CO)_4P(C_6H_5)_3$ yields a *cis–trans* mixture.

Oligomerisation and polymerisation

Several transition metal complexes catalyse the union of a number of unsaturated molecules. The mechanism generally involves combination of insertion, oxidative addition and reductive elimination reactions. Conversions may be specific or yield a variety of products. Thus the $TiCl_4$–$(CH_3)_3Al_2Cl_3$ system gives varying yields of C_4, C_6 and C_8 alkenes from ethene while the $Co(acac)_3$–$Al(C_2H_5)_3$ system gives a 99.5 per cent yield of butenes (95 per cent but-2-ene) from the same substrate. Typical behaviour is shown by oligomerisation of $RC\equiv CH$ using $Ni(CO)_2(P(C_6H_5)_3)_2$. The active species in this reaction does not contain carbon monoxide. An induction period presumably due to slow exchange of carbon monoxide precedes the reaction. The suggested mechanism involves alkyne insertion into a nickel–carbon bond.

246

Cyclotrimerisation is also possible using the same complex, and benzene may be obtained in 88 per cent yield when R=H and 1,2,4-tris(hydroxymethyl)benzene in 65 per cent yield when R=CH$_2$OH.

Metal ion catalysed redox reactions

Reactions of this type are particularly widespread. Theoretically any system involving the transfer of electrons will be subject to this type of catalysis. The metal ion catalyst merely provides a more favourable route for transfer of the electron(s). Such catalytic activity has been exploited to a considerable extent in industrial processes and is of great significance in numerous biological processes. A typical system is represented

$$\underset{\text{oxidant}}{A} + \underset{\text{reductant}}{B} \rightarrow \text{products} \qquad 247$$

catalysed by metal ion M^{n+} *via* the mechanism

$$A + M^{n+} \rightarrow \text{product} + M^{(n+1)+} \qquad 248$$

$$M^{(n+1)+} + B \rightarrow \text{product} + M^{n+} \qquad 249$$

yielding

$$A + B \xrightarrow{M^{n+} \rightleftharpoons M^{(n+1)+}} \text{products} \qquad 250$$

A particular example is the reaction between cerium(IV) and thallium(I)

$$2Ce(\text{IV}) + Tl(\text{I}) \rightarrow Tl(\text{III}) + 2Ce(\text{III}) \qquad 251$$

which is catalysed by manganese(II).
The mechanism

$$Ce(\text{IV}) + Mn(\text{II}) \rightarrow Ce(\text{III}) + Mn(\text{III}) \qquad 252$$

$$Mn(\text{III}) + Ce(\text{IV}) \rightarrow Ce(\text{III}) + Mn(\text{IV}) \qquad 253$$

$$Mn(\text{IV}) + Tl(\text{I}) \rightarrow Mn(\text{II}) + Tl(\text{III}) \qquad 89$$

is suggested.

It is noteworthy that the manganese(II) catalyst is mediating through both one and two electron transfers thereby satisfying the requirement of cerium(IV) for a one electron transfer and thallium(I) for a two electron transfer.

A number of examples the majority of which are relatively easy to demonstrate are shown in Table 21.

An industrially significant example of redox catalysis which admirably illustrates the manner in which the behaviour may be exploited is the oxidation of ethene by aqueous palladium(II) chloride

$$PdCl_4^{2-} + C_2H_4 + H_2O \rightarrow CH_3CHO + Pd + 4Cl^- + 2H^+ \qquad 254$$

Table 21. Some metal ion catalysed redox reactions.[a]

Reaction	Catalyst	Notes
V(III)–Fe(III)	Cu(I) ⇌ Cu(II)	V(III) reduces Cu(II), Fe(III) oxidises Cu(I)
$C_2O_4^{2-}$–Cl_2	Mn(II) ⇌ Mn(IV)	
I^-–$S_2O_8^{2-}$	Cu(I) ⇌ Cu(II)	See p 87
Cr(III)–$S_2O_8^{2-}$	Ag(I) ⇌ Ag(II) ⇌ Ag(III)	
$S_2O_3^{2-}$–Fe(III)	Cu(I) ⇌ Cu(II)	See p 85
HSO_3^-–Fe(III)	Cu(I) ⇌ Cu(II)	Cu(II) oxidises HSO_3^- radicals much faster than Fe(III) (see p 43)
N_2H_4–methylene blue	Mo(V) ⇌ Mo(VI)	
$C_2O_4^{2-}$–Ce(IV)	Os(VI) ⇌ Os(VII) ⇌ Os(VIII)	

[a] See also the redox catalysed substitution discussed on p 36 and the examples mentioned in Table 20.

The suggested mechanism is represented

$$PdCl_4^{2-} + C_2H_4 \longrightarrow PdCl_3C_2H_4^- + Cl^- \qquad 255$$

[structure] $+ H_2O + OH^- \xrightarrow{slow}$ HOCH$_2$CH$_2$—Pd—OH$_2$ + Cl$^-$ 256

[structure with] H$_2$O—Pd—CH$_2$C—OH $\longrightarrow$ 2Cl$^-$ + Pd + CH$_3$CHO + H$_3$O$^+$ 257

This reaction consumes palladium(II) chloride stoichiometrically producing palladium metal; it may, however, be rendered homogeneous by adding copper(II) which oxidises the palladium metal back to palladium(II).

$$Pd + 4Cl^- + 2Cu(II) \rightarrow PdCl_4^{2-} + 2Cu(I) \qquad 258$$

The copper(I) may be reoxidised by oxygen (air)

$$4Cu(I) + O_2 + 4H^+ \rightarrow 4Cu(II) + 2H_2O \qquad 259$$

The overall reaction, the Wacker process, is represented

$$2C_2H_4 + O_2 \xrightarrow[Cu(II)]{PdCl_4^{2-}} 2CH_3CHO \qquad 260$$

Autoxidation

The term autoxidation is generally applied to slow oxidations caused by oxygen (air) at low temperature. The process involves free radicals and is generally accelerated by one electron oxidants (*eg* manganese(III) and cobalt(III)). (The high reactivity of oxygen towards free radicals results from the fact that the oxygen molecule is itself a diradical.) The role of peroxides in the initiation of these reactions is of some significance. Metal ions have little effect on the oxygen uptake of hydrocarbons in the absence of traces of peroxides. Other substrates such as aldehydes or ketones, which may be oxidised directly by manganese(III) and cobalt(III) are, however, autoxidised in the presence of these metal ions although peroxides are initially absent. For the industrially important cobalt catalysed oxidation of a hydrocarbon (RH) the process may be represented

$$Co^{3+} + ROOH \rightarrow Co^{2+} + H^+ + ROO\cdot \quad \text{chain initiation} \quad 261$$
$$Co^{2+} + ROOH \rightarrow Co^{3+} + OH^- + RO\cdot \quad \quad\quad\quad\quad\quad\quad 262$$
$$ROO\cdot + RH \rightarrow ROOH + R\cdot \quad\quad\quad\quad\quad\quad\quad\quad\quad 263$$
$$R\cdot + O_2 \rightarrow ROO\cdot \quad \text{propagation} \quad 264$$
$$RO\cdot + RH \rightarrow ROH + R\cdot \quad\quad\quad\quad\quad\quad\quad\quad\quad 265$$
$$2ROO\cdot \rightarrow \text{products} \quad\quad\quad\quad\quad\quad\quad\quad\quad\quad 266$$
$$2R\cdot \rightarrow \text{products} \quad \text{termination} \quad 267$$
$$R\cdot + ROO\cdot \rightarrow \text{products} \quad\quad\quad\quad\quad\quad\quad\quad 268$$

The reaction is generally initiated by addition of cobalt(II) and a little ROOR, ROOH or H_2O_2 to the hydrocarbon. The napthenate or stearate is used to solubilise the cobalt(II).

Iron(II) catalysed peroxide oxidations

These reactions, though not autoxidations, are conveniently discussed here. The usefulness of iron(II) as a catalyst for peroxide oxidations in acid solution was discovered by H. J. H. Fenton in 1894. The iron(II)–hydrogen peroxide reagent is now termed Fenton's reagent.

It is clear that it is the hydroxyl radical formed *via*

$$Fe(II) + H_2O_2 \rightarrow Fe(OH)^{2+} + \cdot OH \quad\quad\quad 269$$

which is responsible for oxidation of the organic species. Typical examples are the oxidation of ethanol and benzene. The oxidation of ethanol is a chain reaction requiring only a small quantity of iron(II). The mechanism is represented

$$Fe(II) + H_2O_2 \rightarrow Fe(OH)^{2+} + \cdot OH \quad\quad\quad 269$$
$$\cdot OH + CH_3CH_2OH \rightarrow H_2O + CH_3\dot{C}HOH \quad\quad\quad 270$$
$$FeOH^{2+} + CH_3\dot{C}HOH \rightarrow CH_3CHO + H_2O + Fe(II) \quad\quad\quad 271$$

The oxidation of benzene may lead to phenol or biphenyl.

$$Fe(II) + H_2O_2 \longrightarrow Fe(OH)^{2+} + {}^{\cdot}OH \qquad 269$$

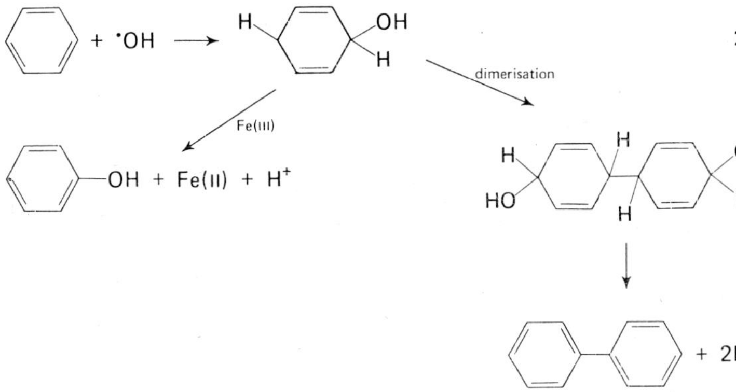

$$\text{[benzene]} + {}^{\cdot}OH \longrightarrow \text{[cyclohexadienyl-OH]} \qquad 272$$

dimerisation

[phenol] + Fe(II) + H⁺ ← Fe(III)

[dihydroxy-biphenyl intermediate]

↓

[biphenyl] + 2H₂O

Experimental

The copper(II) assisted hydrolysis of ethylaminoethanoate (p 79)

Requirements

$C_2H_5CO_2CH_2NH_2 \cdot HCl$
$NH_2C(CH_2OH)_3$
$Cu(NO_3)_2 \cdot 3H_2O \quad 10^{-1}$ M (24.16 g dm^{-3})
HCl $\quad 10^{-1}$ M
NaOH $\quad 10^{-1}$ M (4.00 g dm^{-3})
pH meter.

Kinetic investigation (25–35 °C)

To 0.56 g ethylaminoethanoate hydrochloride in a 250 cm³ flask add 4.2 g tris(hydroxymethyl)aminomethane (tris buffer), 27 cm³ 10^{-1} M sodium hydroxide and 25 cm³ 10^{-1} M copper(II). Make up to 250 cm³. Titrate 25 cm³ aliquots *vs* time, to a pH of 4.0, with 10^{-1} M hydrochloric acid. Samples are conveniently examined at 10–15 min intervals. The effect of copper(II) concentration may be examined. Typical results are shown in *Fig 10*.

The copper(II) catalysed iron(III)–trioxothiosulphate(VI) (thiosulphate) reaction

Requirements

$Na_2S_2O_3 \cdot 5H_2O \quad 10^{-1}$ M (24.8 g dm^{-3})
$NH_4Fe(SO_4)_2 \cdot 12H_2O \quad 5 \times 10^{-3}$ M (2.41 g dm^{-3})
$CuSO_4 \cdot 5H_2O \quad 10^{-3}$ M (0.25 g dm^{-3}).

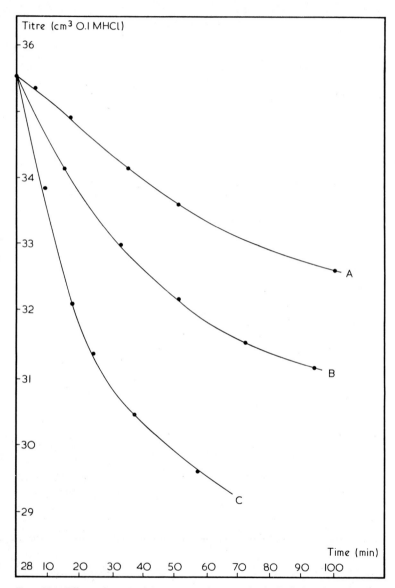

Fig. 10. Effect of copper(II) concentration on the hydrolysis of ethylaminoethanoate. Conditions as in text. $T = 27\,°C$. (A) 5×10^{-3} M Cu(II). (B) 10^{-2} M Cu(II). (C) 2×10^{-2} M Cu(II).

METAL ION ASSISTED PROCESSES

Demonstration (20–25 °C)

To 10 cm^3 10^{-1} M sodium trioxothiosulphate(VI) add 10 cm^3 5×10^{-3} M iron(III). The solution turns instantly brown–purple (Fe(S$_2$O$_3$)$_n^{3-2n}$, p 5) and then slowly colourless over 2–3 min.

$$2Fe^{3+} + 2S_2O_3^{2-} \rightarrow S_4O_6^{2-} + 2Fe^{2+} \qquad 273$$

A suggested mechanism for the catalysis is

$$2(Cu(II) + S_2O_3^{2-} \rightarrow S_2O_3^{\cdot} + Cu(I)) \qquad 274$$

$$2S_2O_3^{\cdot} \rightarrow S_4O_6^{2-} \qquad 275$$

$$2(Cu(I) + Fe(III) \rightarrow Fe(II) + Cu(II)) \qquad 276$$

The copper(II) catalysed peroxodisulphate(VI) oxidation of iodide ion

$$S_2O_8^{2-} + 2I^- \rightarrow I_2 + 2SO_4^{2-} \qquad 277$$

This system is conveniently examined by conversion to a clock reaction *via* addition of trioxothiosulphate(VI) ion (p 72).[1]

Requirements
KI   2×10^{-1} M (33.2 g dm^{-3})
Na$_2$S$_2$O$_3$.5H$_2$O   10^{-2} M (2.48 g dm^{-3})
(NH$_4$)$_2$S$_2$O$_8$ 10^{-1} M (22.8 g dm^{-3})
CuSO$_4$.5H$_2$O   10^{-1} M (24.97 g dm^{-3})
1 per cent starch.

Demonstration

To a solution of 10 cm^3 2×10^{-1} M potassium iodide, 10 cm^3 10^{-2} M sodium trioxothiosulphate(VI) and a few drops of starch add 20 cm^3 10^{-1} M freshly prepared peroxodisulphate(VI) solution. Note the time for the blue starch–iodine colour to appear.

Repeat but add 0.5 cm^3 10^{-1} M copper(II) to the peroxodisulphate(VI) solution. Varying concentrations of copper(II) may be examined. The reaction may be used analytically for the determination of low concentrations of copper(II) (*Fig. 11*).

The mechanism of the reaction is not altogether clear. The peroxodisulphate(VI)–iodide reaction (absence of catalyst) may easily be shown to obey the rate equation

$$\frac{d[I_2]}{dt} = k[S_2O_8^{2-}][I^-]$$

The suggested mechanism is

$$S_2O_8^{2-} + I^- \rightarrow S_2O_8I^{3-} \quad \text{rate determining} \qquad 278$$

$$S_2O_8I^{3-} + I^- \rightarrow 2SO_4^{2-} + I_2 \qquad 279$$

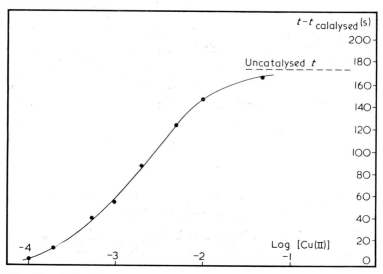

FIG. 11. Calibration curve for the determination of copper(II) via catalysis of reaction 277. Conditions as in text. $T = 20\,°C$.

The reaction mixture is, however, capable of polymerising cyanoethene (acrylonitrile) which is indicative of the presence of free radicals. Both $SO_4^-\cdot$ and $I\cdot$ are possibilities but it is not clear if they arise via $S_2O_8I^{3-}$ or not.

Various mechanisms for the catalysis may be suggested. Iron(III) is also a catalyst for the reaction. The relevant electrode potentials (Table 22) indicate that iron(III) is capable of oxidising iodide to iodine and that peroxodisulphate(VI) is capable of oxidising iron(II) to iron(III). The copper(II) oxidation, despite a negative $E^\ominus$ value, is driven to the right as a result of the insolubility of copper(I) iodide. It would also be favoured by removal of copper(I) via oxidation.

A redox catalysis

$$2Cu(II) + 2I^- \rightarrow 2Cu(I) + I_2 \qquad\qquad 280$$

$$2Cu(I) + S_2O_8^{2-} \rightarrow 2Cu(II) + 2SO_4^{2-} \qquad\qquad 281$$

may thus be suggested.

Table 22. Electrode potentials.

Redox couple	$E^\ominus$ (V)
$Cu(I) \rightarrow Cu(II) + e^-$	-0.153
$Fe(II) \rightarrow Fe(III) + e^-$	-0.77
$2I^- \rightarrow I_2 + 2e^-$	-0.535
$2SO_4^{2-} \rightarrow S_2O_8^{2-} + 2e^-$	-2.01

The copper(II) oxidation of iodide ion may be shown to be rapid. The iron(III) oxidation is much slower (p 57). The catalysis is less marked in the case of iron(III). The peroxodisulphate(VI) oxidation of iron(II) may be shown to be rapid. The oxidation of copper(I) is somewhat slower. An alternative mechanism involves coordination of peroxodisulphate(VI) and no change in oxidation state of the metal ion catalyst.

$$S_2O_8^{2-} + M^{n+} \rightleftharpoons MS_2O_8^{(n-2)+} \qquad 282$$

$$MS_2O_8^{(n-2)+} + I^- \rightarrow I^+ + 2SO_4^{2-} + M^{n+} \quad (I^+ \text{ considered as HIO}) \qquad 283$$

$$I^+ + I^- \rightarrow I_2 \qquad 284$$

The species $FeS_2O_8^+$ has been characterised in the iron(III) catalysed reaction.

The chromium(III)–edtaH$_2^{2-}$ reaction

This ligand exchange reaction described on p 28 is catalysed by anions such as carbonate, dioxonitrate(III) and trioxosulphate(IV). Addition of 2 cm^3 of a 10^{-1} M solution of these ions to the reaction mixture (p 28) enables demonstration. The mechanism is not clear. There is, however, a considerable change in the visible spectrum of chromium(III) on adding the catalyst. The process may proceed *via* complex formation involving the catalyst, *eg*

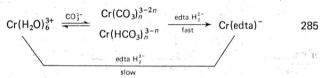

285

The carbonate complexes are probably outer sphere, serving to labilise the coordinated water molecules.

Reference
1. Nuffield Advanced Science, Chemistry, *Students Book II*, Expt 16.3. Harmondsworth: Penguin, 1970.

7. Oscillating Reactions

The first report of an oscillating reaction, namely that of oscillations in passivating electrodes, dates back to 1828. Since then numerous electrochemical examples involving both current oscillations under potentiostatic conditions and potential oscillations under galvanostatic conditions have been reported. Examples include iron in trioxonitric(V) acid and the reduction of hydrogentetraoxochromate(VI) on mercury. Other heterogeneous or inhomogeneous periodic phenomena are represented by the periodic precipitation patterns known as Liesegang rings,[1,2] the catalytic decomposition of hydrogen peroxide by mercury in weakly alkaline solution and the rate of dissolution of some metal–metal amalgams in water or acid. Oscillations which are apparently driven by temperature changes are provided by the gas phase oxidation of hydrocarbons at low ($\leqslant 400\,^\circ$C) temperature.

The study of homogeneous* liquid phase oscillating systems is, however, more recent, and though the first report appeared in 1921[3] considerable interest was only aroused during the 1960s. Easily observable examples are the Belousov–Zhabotinskii reaction (the cerium(IV) catalysed trioxobromate(V) oxidation of propanedioic acid) and the trioxoiodate(V)–hydrogen peroxide–manganese(II)–propanone reaction. Biochemical examples are numerous. Oscillations have been observed in the glycolytic pathway, the synthesis of certain proteins and in some peroxidase oxidations.

This chapter concentrates on the liquid phase systems, discusses the requirements for observing these phenomena and examines, in detail, the Belousov–Zhabotkinskii reaction which is as yet the only well understood system of this type.

In any homogeneous system at constant temperature and pressure all spontaneous chemical changes must be accompanied by a decrease in the free energy of the system. Oscillations about the final equilibrium state are impossible. The oscillations observed are thus about a steady state *en route* to equilibrium. In the examples known at present, oscillations are only observed under conditions far from equilibrium. It therefore follows that it is the intermediates which are subject to oscillatory behaviour. The reactants are consumed in a stepwise process and the products produced similarly.

---

* The reactions discovered so far are affected to some extent by stirring and so cannot strictly be considered homogeneous. They are, however, generally considered to be homogeneous in stirred solution.

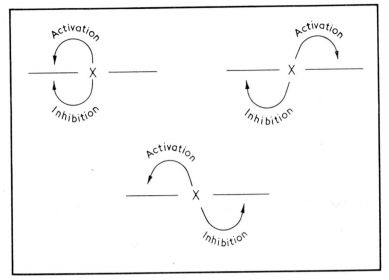

FIG. 12. Some possible feedback mechanisms. A delay is required in order to observe oscillations.

The principal requirement for observing oscillations is feedback. The oscillations are the result of two counteracting feedback mechanisms. 'Positive feedback' leading to propagation and 'negative feedback' leading to recovery and overshoot phenomena. Several possibilities may be suggested for the feedback (*Fig. 12*). For instance (*Fig. 12*), if the effect due to the catalytic feedback (propagation of X) is equal to that due to the inhibitory feedback (destruction of X) a steady state is obtained and no oscillations are observed. If, however, a delay is introduced into one of the loops the steady state may be unstable. Expansion of the hypothetical reaction to give

$$A \to X \to B \to C \to \qquad 286$$

such that A→X is catalysed (*ie* is autocatalytic) and inhibited by B may introduce a delay of the type required, provided the reactions X→B and B→C proceed at suitable rates. The reaction A→X will proceed until the concentration of B is sufficient to inhibit it. At this point, as the concentration of X can still be high, it is possible that the concentration of B may continue to rise even though X is no longer being produced. This can then result in an overshoot in the concentration of B. Eventually X becomes depleted with B being removed to a sink of products (→C→), the concentration of B will fall. This will continue until the reaction A→X is no longer inhibited. The process may then be repeated.

The Belousov–Zhabotinskii reaction

The cerium(III) catalysed oxidation of propanedioic acid by trioxobromate(V) ion. In this reaction, in stirred solution, oscillations in cerium(IV) and bromide ion concentrations are observed. The evolution of carbon dioxide, production of 2-bromopropanedioic acid and the temperature increase in a stepwise manner. The behaviour is shown schematically in *Fig. 13* and is conveniently discussed by considering the system to be composed of the three gross reactions

$$BrO_3^- + 2Br^- + 3CH_2(CO_2H)_2 + 3H^+ \rightarrow 3BrCH(CO_2H)_2 + 3H_2O \qquad 287$$

$$BrO_3^- + 4Ce(III) + CH_2(CO_2H)_2 + 5H^+ \rightarrow \\ BrCH(CO_2H)_2 + 4Ce(IV) + 3H_2O \qquad 288$$

$$4Ce(IV) + BrCH(CO_2H)_2 + 2H_2O \rightarrow \\ Br^- + 4Ce(III) + HCO_2H + 2CO_2 + 5H^+ \qquad 289$$

FIG. 13. Schematic representation of the Belousov–Zhabotinskii reaction.

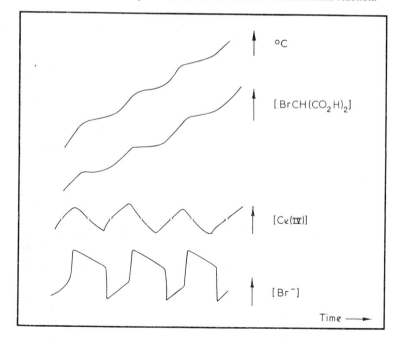

287 is considered to be composed of the production of bromine and subsequent bromination of propanedioic acid.

$$BrO^- + 5Br^- + 6H^+ \rightarrow 3Br_2 + 3H_2O \qquad 290$$

$$Br_2 + CH_2(CO_2H)_2 \rightarrow BrCH(CO_2H)_2 + Br^- + H^+ \qquad 291$$

The mechanism of 290 has been shown to be

$$BrO_3^- + Br^- + 2H^+ \rightarrow HBrO_2 + HBrO \qquad 292$$

$$HBrO_2 + Br^- + H^+ \rightarrow 2HBrO \qquad 293$$

$$HBrO + Br^- + H^+ \rightleftharpoons Br_2 + H_2O \qquad 294$$

The bromination 291 proceeds via an enolisation mechanism

$$CH_2(CO_2H)_2 \rightleftharpoons (OH)_2C{=}CHCO_2H \qquad 295$$

$$(OH)_2C{=}CHCO_2H + Br_2 \rightarrow BrCH(CO_2H)_2 + H^+ + Br^- \qquad 296$$

Further bromination to 2,2-dibromopropanedioic acid may take place. 288 is considered to be composed of the trioxobromate(V) oxidation of cerium(III) which is also a source of bromine and is hence accompanied by the bromination 291.

It is the behaviour of this trioxobromate(V) oxidation which is crucial to the production of oscillations. Cerium(III) is not oxidised directly by trioxobromate(V) but *via* bromine dioxide produced by reaction of dioxobromic(III) acid formed in 292 and trioxobromate(V).

$$HBrO_2 + BrO_3^- + H^+ \rightarrow 2BrO_2^{\cdot} + H_2O \qquad 297$$

$$BrO_2^{\cdot} + Ce(III) + H^+ \rightarrow HBrO_2 + Ce(IV) \qquad 298$$

The immediate consequence of this reaction is that dioxobromic(III) acid is produced autocatalytically. Two factors exert a control on this build-up of dioxobromic(III) acid.

1. Disproportionation

$$2HBrO_2 \rightarrow HBrO_3 + HBrO \qquad 299$$

2. Competition between trioxobromate(V) and bromide ion for the dioxobromic(III) acid *via* 293 and 297. In practice, the competition is very great and at a concentration of $>ca\ 10^{-6}$ M bromide ion reaction 297 is effectively inhibited. Thus the oxidation of cerium(III) is autocatalytic but inhibited by bromide ion. The gross reaction 289 is the inevitable consequence of the presence of both cerium(IV) and 2-bromopropanedioic acid. This reaction provides the bromide which eventually inhibits the production of cerium(IV). It should also be noted that cerium(IV) may oxidise propanedioic acid.

$$6Ce(IV) + CH_2(CO_2H)_2 + 2H_2O \rightarrow HCO_2H + 2CO_2 + 6Ce(III) + 6H^+ \qquad 300$$

With this background information it is now possible to consider what happens when the reactants trioxobromate(v)–tetraoxosulphuric(vi) acid–propanedioic acid and cerium(iii) or (iv) are mixed. The initial step, irrespective of whether or not bromide has been added, is the decrease in concentration of bromide ion. Conditions are now such that cerium(iii) may be oxidised to cerium(iv). If the cerium is initially present as cerium(iv) this is reduced to cerium(iii) *via* oxidation of propanedioic acid (300) subsequent to reoxidation. The oxidation of cerium(iii) leads to the autocatalytic production of dioxobromic(iii) acid (298) and eventually to a source of bromine. Bromination of the propanedioic acid is now proceeding and the reaction is in an induction period prior to observation of the oscillations. During this period 2-bromopropanedioic acid is accumulating and cerium(iv) is oxidising propanedioic acid and 2-bromopropanedioic acid as available. Eventually the bromide ion concentration produced *via* the cerium(iv) oxidation 289 and the oxobromic(i) acid oxidation of methanoic acid (itself inert to cerium(iv)) (301) is high enough to inhibit the oxidation of cerium(iii).

$$HCO_2H + HBrO \rightarrow CO_2 + H^+ + H_2O + Br^-$$ 301

With cerium(iv) no longer being produced the cerium(iv) concentration now falls as the oxidation of propanedioic acid and 2-bromopropanedioic acid proceeds. Bromide ion is removed *via* the trioxobromate(v)–bromide reaction 290 and subsequent bromination of propanedioic acid 291. Thus the inevitable consequence of the decrease in cerium(iv) concentration is that the bromide concentration will also decrease. As this reaches the point where the oxidation of cerium(iii) is no longer inhibited, oxidation occurs with autocatalytic production of dioxobromic(iii) acid which removes the remaining bromide ion *via* 293 thus providing the overshoot phenomena. With the cerium(iv) oxidation again proceeding the process may thus be repeated. The feedback mechanism may be represented

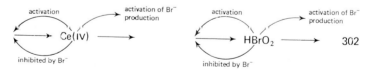

The overall reaction may be considered as transferring trioxobromate(v) and propanedioic acid from a 'sink of reactants' to brominated organic species and carbon dioxide as a 'sink of products'. The mechanism is summarised in Scheme 7.

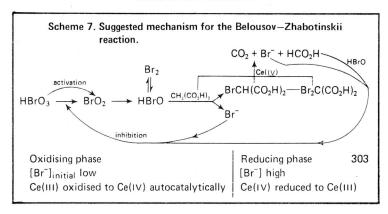

Scheme 7. Suggested mechanism for the Belousov–Zhabotinskii reaction.

Oxidising phase
$[Br^-]_{initial}$ low
Ce(III) oxidised to Ce(IV) autocatalytically

Reducing phase
$[Br^-]$ high
Ce(IV) reduced to Ce(III)

303

Experimental

The reactions described in this section are best observed in stirred solution. An investigation may be colorimetric or potentiometric. The potentiometric method using a standard calomel electrode as reference (*Fig. 14*) is particularly useful. The electrode potential may be displayed on a millivoltmeter or a potentiometric recorder.

The cerium(IV) catalysed trioxobromate(V) oxidation of propanedioic acid

Requirements

$CH_2(CO_2H)_2$ 5×10^{-1} M (52.03 g dm^{-3})
$KBrO_3$ 2.5×10^{-1} M (41.75 g dm^{-3})
$(NH_4)_2Ce(NO_3)_6$ 10^{-2} M in 6M H_2SO_4 (5.48 g dm^{-3} 6 M H_2SO_4).

Demonstration

To 20 cm^3 5×10^{-1} M propanedioic acid and 10 cm^3 2.5×10^{-1} M potassium trioxobromate(V) add 10 cm^3 of a solution 10^{-2} M with respect to cerium(IV) and 6 M with respect to tetraoxosulphuric(VI)

FIG. 14. Cell for observation of the redox behaviour of oscillating reactions.

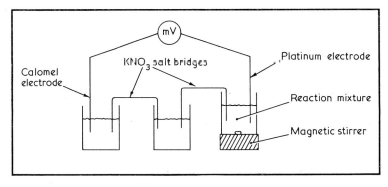

acid. The oscillations are from yellow (Ce(IV)) to colourless (Ce(III)). The tris (1,10-phenanthroline) iron(II) ion (Ferroin) may be used as indicator in which case the oscillations are from blue (Ce(IV)/Fe(III)) to red (Ce(III)/Fe(II)).

Many variations on this reaction are possible. A useful alternative is to replace cerium(IV) with manganese(II) and propanedioic acid with 2-hydroxybutanedioic (malic) acid.

Requirements

HOCOCH$_2$CHOHCO$_2$H   1 M (134.09 g dm^{-3})
KBrO$_3$ 2.5×10^{-1} M (41.75 g dm^{-3})
MnSO$_4$.4H$_2$O 10^{-2} M in 6 M H$_2$SO$_4$ (2.23 g dm^{-3} 6 M H$_2$SO$_4$).

Demonstration

To 20 cm^3 1 M 2-hydroxybutanedioic acid and 10 cm^3 2.5×10^{-1} M potassium trioxobromate(V) add 10 cm^3 of a solution 10^{-2} M with respect to manganese(II) and 6 M with respect to tetraoxosulphuric(VI) acid. The oscillations are from yellow (Mn(III)-2-hydroxybutanoate complex) to colourless (Mn(II)).

The manganese(II)-trioxoiodate(V)-hydrogen peroxide-propanone reaction

Requirements

(CH$_3$)$_2$CO
H$_2$SO$_4$ 1 M
H$_2$O$_2$ '100 volume'
KIO$_3$ 10^{-1} M (21.4 g dm^{-3})
MnSO$_4$.4H$_2$O 7.5×10^{-2} M (16.73 g dm^{-3}).

Demonstration

To 20 cm^3 propanone add 5 cm^3 1 M tetraoxosulphuric(VI) acid, 20 cm^3 10^{-1} M potassium trioxoiodate(V), 5 cm^3 '100 volume' hydrogen peroxide and 5 cm^3 7.5×10^{-2} M manganese(II). The oscillations (*Fig. 15*) are from yellow (iodine) to colourless.

In view of the hazards associated with ketone–peroxide mixtures solutions should not be stored, distilled or allowed to evaporate to dryness.

The detailed mechanism of this reaction has not yet been elucidated. An outline suitable for discussion is shown below.

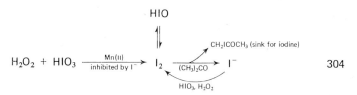

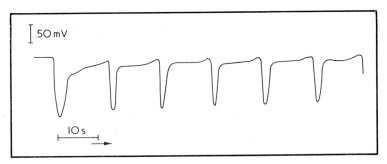

FIG. 15. Typical redox potential oscillations (Pt vs SCE). Conditions as in text. $T = 25\,°C$.

An experiment to examine the mechanism of the iodination of propanone is available.[4] The mechanism of the trioxoiodate(v)–iodide reaction is discussed on p 68.

References
1. J. Liesegang, *Naturw. Wschr.*, 1898, **11**, 353.
2. C. L. Strong, *Scient. Am.*, 1969, **220** (6), 130.
3. W. C. Bray, *J. Am. chem. Soc.*, 1921, **43**, 1262.
4. Nuffield Advanced Science, Chemistry, *Students Book II*, Expt 14.2. Harmondsworth: Penguin, 1970.

Appendix 1: Preparations

The experimental work described in this monograph requires the use of several compounds which are not generally available. The preparation, analysis and in some cases thermal decomposition together with the kinetic investigation(s) can form useful 'integrated exercises'. The compounds and some useful references are given in Table 23.

Table 23. Compounds for experiments.

Compound	References
cis and trans [Co(en)$_2$Cl$_2$]Cl	This appendix
Co(en)$_3$Cl$_3$	2
Co(NH$_3$)$_6$Cl$_3$ [a]	1, 2, 3,
K$_3$Cr(C$_2$O$_4$)$_3$.3H$_2$O	2
K$_3$Fe(C$_2$O$_4$)$_3$.3H$_2$O [a]	2, 3
K$_3$Co(C$_2$O$_4$)$_3$.3 5H$_2$O	2, 3, 4
K$_2$(Cu(C$_2$O$_4$)$_2$.2H$_2$O	5

[a] These compounds are available commercially.
1. W. G. Palmer, *Experimental inorganic chemistry*. Cambridge: CUP, 1970.
2. G. G. Schlessinger, *Inorganic laboratory preparations*. New York: Chemical Publishing, 1962.
3. D. Abbott, *Practical inorganic chemistry for sixth forms*. London: Dent, 1966.
4. P. W. Wiggans, *Educ. Chem.* 1975, **12**, 54.
5. J. R. Darley and J. I. Hoppé, *J. chem. Educ.* 1972, **49**, 365.

Preparation of trans-dichlorobis(ethane-1,2-diamine) cobalt(III) chloride ([Co(en)$_2$Cl$_2$]Cl)

To a solution of cobalt(II) chloride-6-water (16 g) in water (50 cm^3) add 60 g 10 per cent ethane-1,2-diamine. Draw air (filter pump) through the resultant mixture for *ca* 3 h. Add 35 cm^3 concentrated hydrochloric acid to the oxidised solution and evaporate on a steam bath (fume cupboard) until a crust forms on the surface. Allow to stand overnight. Filter. Dark green [Co(en)$_2$Cl$_2$]Cl. 2H$_2$O.HCl is obtained. (The structure of this acid adduct is *trans*[Co(en)$_2$Cl$_2$]$^+$.[H$_5$O$_2$]$^+$.2Cl$^-$.) Grind the product with *ca* 20 cm^3 methanol (removes hydrogen chloride and water). Filter. Dry at 110 °C (removes remaining hydrogen chloride and water). Yield *ca* 8 g.

The preparation may usefully be scaled upwards. 10×, 6 h oxidation gives a yield of *ca* 90 g. Evaporation (steam bath) of a neutral solution of the product yields the violet *cis* isomer.

Suggestions for Further Reading

F. Basolo and R. G. Pearson, *Mechanisms of inorganic reactions*, 2nd edn. New York: Wiley, 1967.

D. Benson, *Mechanisms of inorganic reactions in solution*. New York: McGraw-Hill, 1968.

H. Taube, *Electron transfer reactions of complex ions in solution*. New York: Academic, 1970.

R. G. Wilkins, *The study of kinetics and mechanism of reactions of transition metal complexes*. Boston: Allyn and Bacon, 1974.